Ready Notes

for use with

Physical Geology

Eighth Edition

Charles C. Plummer
California State University at Sacramento

David McGeary
Emeritus of California State University at Sacramento

Diane H. Carlson
California State University at Sacramento

The McGraw-Hill Companies, Inc
Primis Custom Publishing

New York St. Louis San Francisco Auckland Bogota
Caracas Lisbon London Madrid Mexico Milan Montreal
New Delhi Paris San Juan Singapore Sydney Tokyo Toronto

McGraw-Hill 
A Division of The McGraw-Hill Companies

McGraw-Hill's College Custom Series consists of products that are produced from camera-ready copy. Peer review, class testing, and accuracy are primarily the responsibility of the author(s).

2 3 4 5 6 7 8 9 0 QPD 0 9

ISBN 0-07-234020-7

Welcome to

READY NOTES

Your life just got easier! This booklet includes *Ready Notes* to accompany your textbook. *Ready Notes* were designed as a classroom supplement to accompany *Ready Shows*. More importantly, *Ready Notes* were developed for you, the student.

Somewhere in your educational experience, you have undoubtedly encountered a common dilemma facing many students; the feeling of helplessness that comes from trying to write down everything your instructor says and, at the same time, actually paying attention to what is being taught. *Ready Notes* addresses this problem by providing pre-prepared lecture outlines to accompany the *Ready Shows* your instructor will be using in class. Rather than spending time copying material that is already in the book, you will be able to focus on the most important aspects of what your instructor is actually saying. You will still be expected to take notes, but the nature of those notes will change.

Each page in *Ready Notes* includes reproductions of some of the actual projected screens that you will be seeing in class. The *Ready Notes* booklet includes the information for many of the examples that your instructor will be presenting.

It is your responsibility to attend class regularly and to be prepared for class. However, used properly, *Ready Notes* will help you to achieve your goals for the course. Good luck!

CONTENTS

Chapter 1

Physical Geology

eighth edition

Plummer/McGeary/Carlson

INTRODUCTION TO PHYSICAL GEOLOGY

Chapter 1

Who Needs Geology

Understanding our dynamic planet through ***Physical Geology***

Who Needs Geology

- Avoiding Geologic Hazards
  - Earthquakes
  - Volcanoes
  - Other geologic hazards

Who Needs Geology

- Supplying Things We Need
  - Energy
  - Metals
  - Other
- Protecting the Environment
- Understanding Our Surroundings
  - Appreciating scenery while traveling

Overview of Geology--Important Concepts

- Internal Heat Engine--Internal Processes
  - Hot rock from earth's interior flows upward
- External Heat Engine--External Processes

Earth's Interior

- Core
  - liquid outer core; solid inner core
- Mantle
  - most of earth
  - rock
- Crust
  - Oceanic Crust-- denser, thinner
  - Continental Crust-- lighter, thicker

Earth's Interior

- Lithosphere
  - Crust + Uppermost Mantle
  - Rigid (tectonic plates)
- Asthenosphere
  - Beneath lithosphere
  - Mantle
  - Soft-- near the melting point
- Tectonic forces
  - due to movement within the mantle

Theory of Plate Tectonics

- ***Plates*** in motion
- ***Divergent Boundaries***
  - Mid-oceanic ridges
  - **Magma** enters fissures
  - Lithosphere moves away from boundary
- ***Transform Boundaries***
  - Plates slide past one another

Theory of Plate Tectonics

- ***Convergent Boundaries***
  - **Subduction Zone**
  - Magma created at depth
    - Moves upward, solidifies into **igneous rock**
  - **Metamorphic rock**--changed rock that doesn't melt

Scientific Method

- Question raised or problem presented
- Data gathered
- **Hypotheses** proposed
- Predictions made
- Predictions tested
- hypothesis that withstands testing becomes a **theory**

Surficial Processes

- Driven by solar power & gravity
- **Erosion**--due to water, ice, wind, gravity
- Rock formed at high temperature becomes unstable at surface
  - Form new material stable under conditions at earth's surface
  - **equilibrium**
- **Sediment**
  - can solidify into **sedimentary rock**

GEOLOGIC TIME

- Earth is around 4.5 billion years old
- Major subdivisions of geologic time
  - Cenozoic Era- youngest
  - Mesozoic Era- middle (dinosaurs lived then)
  - Paleozoic
    - Began around 545 million years ago
    - Oldest abundant fossils
  - PRECAMBRIAN- all time before Paleozoic

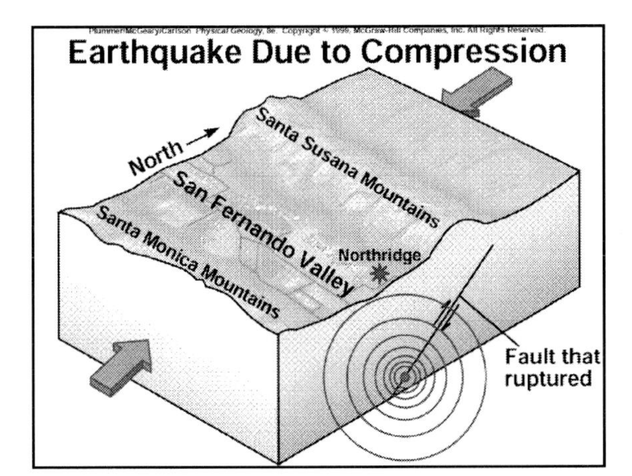

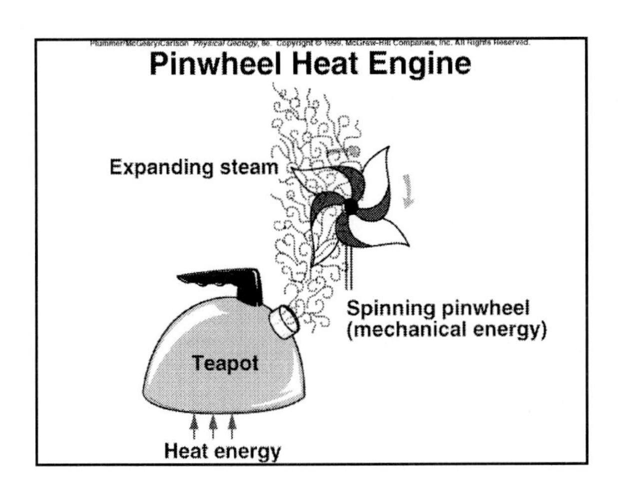

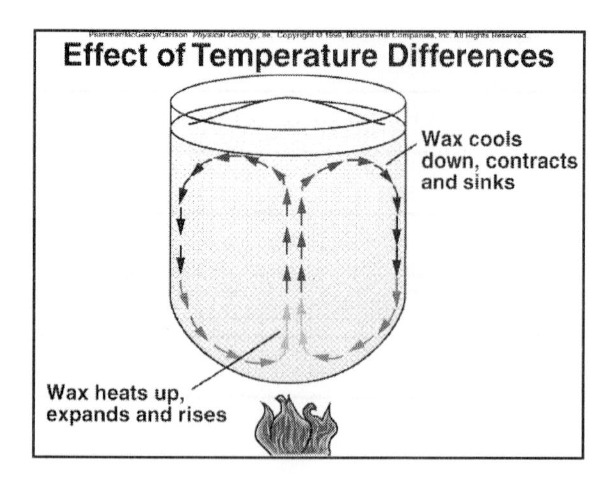
Effect of Temperature Differences
Wax cools down, contracts and sinks
Wax heats up, expands and rises

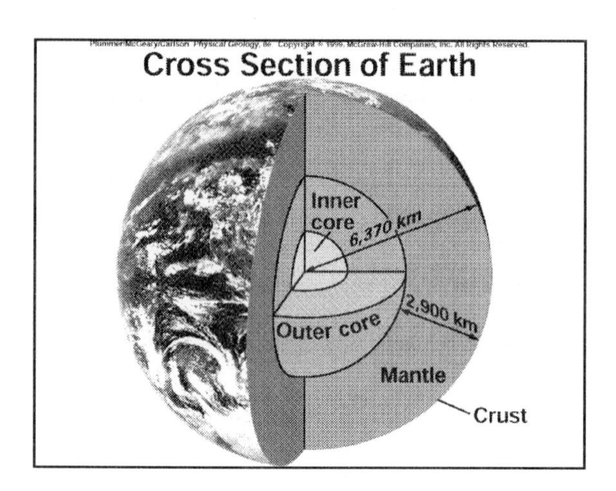
Cross Section of Earth
Inner core
6,370 km
2,900 km
Outer core
Mantle
Crust

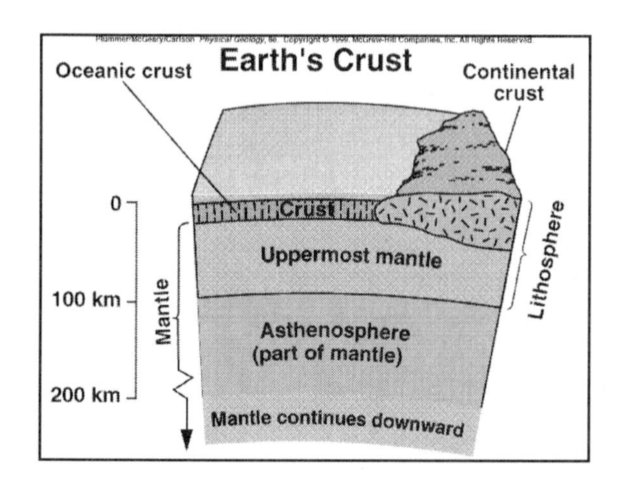
Oceanic crust
Earth's Crust
Continental crust
0
Crust
Uppermost mantle
Lithosphere
100 km
Mantle
Asthenosphere (part of mantle)
200 km
Mantle continues downward

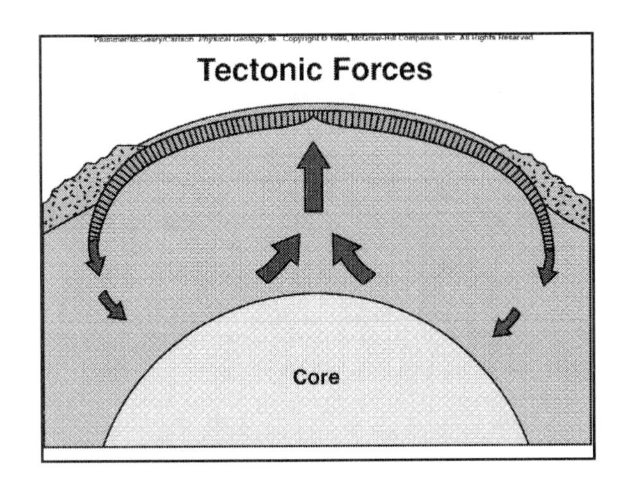
Tectonic Forces
Core

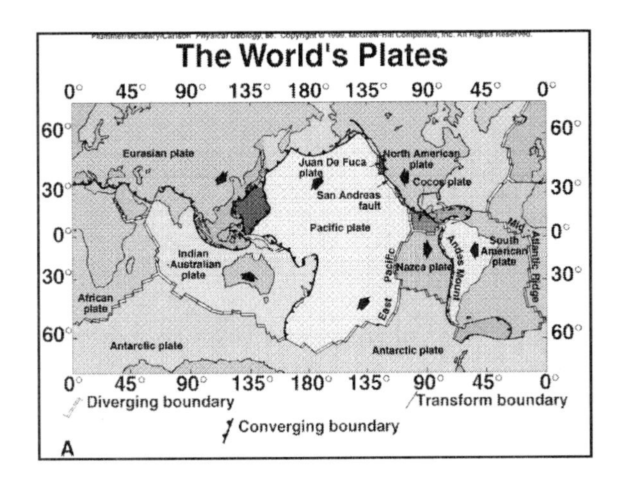
The World's Plates
Eurasian plate
Juan De Fuca plate
North American plate
Cocos plate
San Andreas fault
Pacific plate
Indian -Australian plate
African plate
Antarctic plate
Nazca plate
South American plate
Andes Mount
East Pacific
Mid Atlantic Ridge
Diverging boundary
Transform boundary
Converging boundary
A

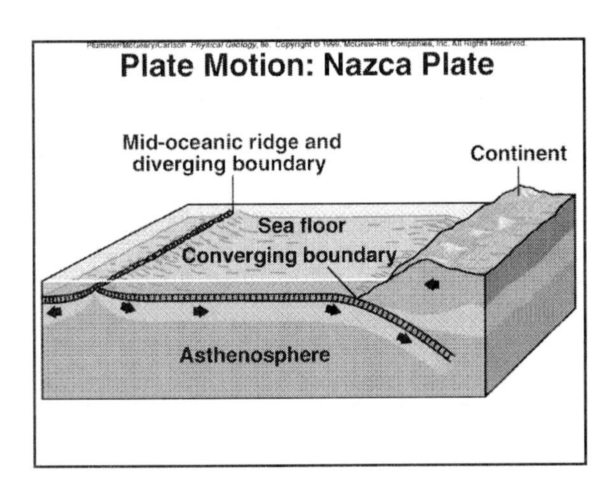
Plate Motion: Nazca Plate
Mid-oceanic ridge and diverging boundary
Continent
Sea floor
Converging boundary
Asthenosphere

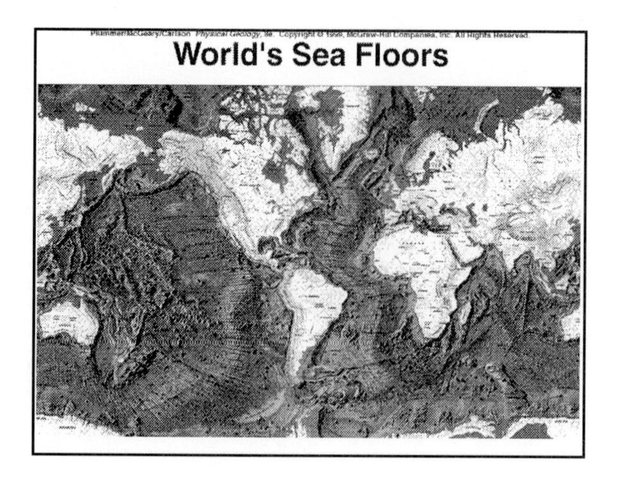
World's Sea Floors

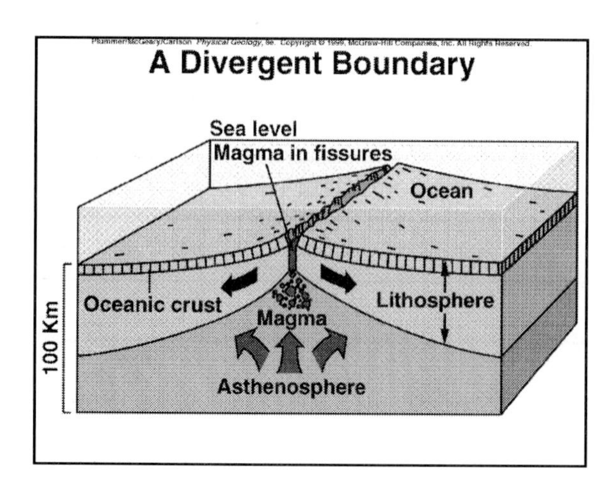
A Divergent Boundary
Sea level
Magma in fissures
Ocean
Oceanic crust
Lithosphere
Magma
Asthenosphere
100 Km

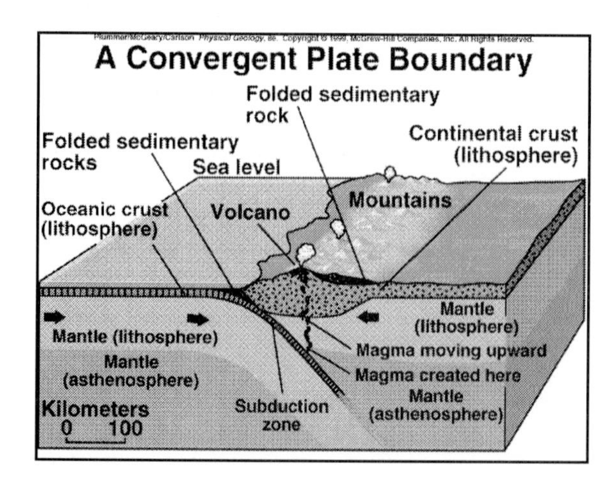
A Convergent Plate Boundary
Folded sedimentary rock
Folded sedimentary rocks
Continental crust (lithosphere)
Sea level
Oceanic crust (lithosphere)
Volcano
Mountains
Mantle (lithosphere)
Mantle (lithosphere)
Mantle (asthenosphere)
Magma moving upward
Magma created here
Mantle (asthenosphere)
Kilometers
0 100
Subduction zone

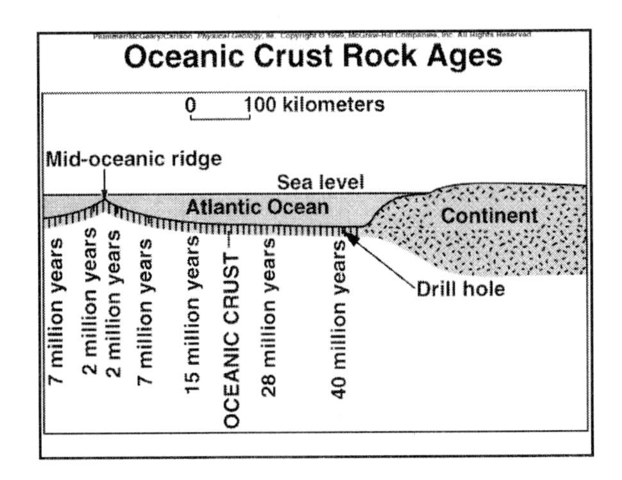
Plummer/McGeary/Carlson Physical Geology, 8e. Copyright © 1999, McGraw-Hill Companies, Inc. All Rights Reserved
Oceanic Crust Rock Ages
0 100 kilometers
Mid-oceanic ridge
Sea level
Atlantic Ocean
Continent
Drill hole
7 million years
2 million years
2 million years
7 million years
15 million years
OCEANIC CRUST
28 million years
40 million years

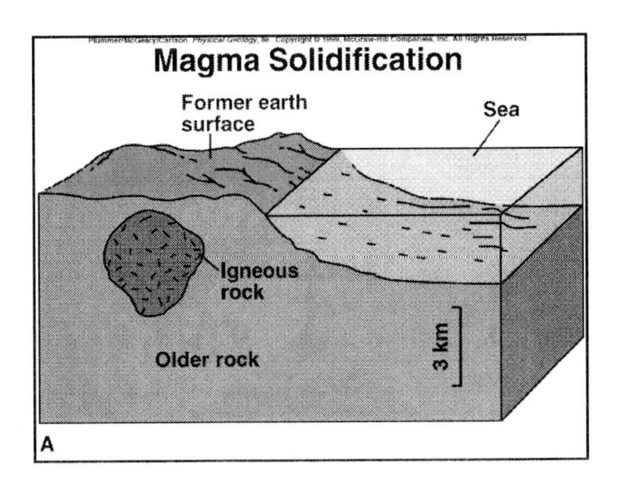
Magma Solidification
Former earth surface
Sea
Igneous rock
Older rock
3 km
A

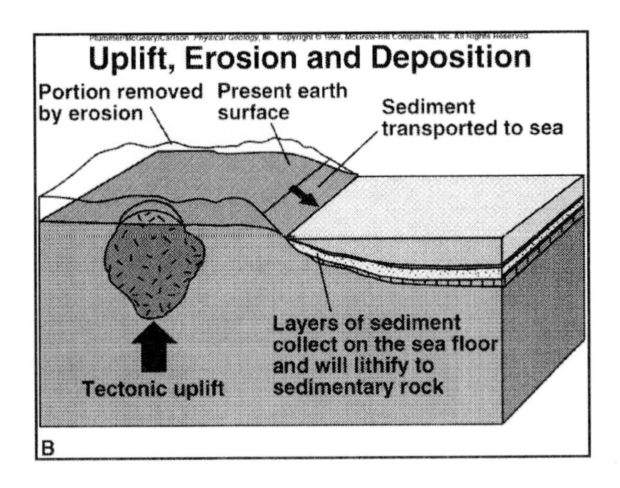
Uplift, Erosion and Deposition
Portion removed by erosion
Present earth surface
Sediment transported to sea
Layers of sediment collect on the sea floor and will lithify to sedimentary rock
Tectonic uplift
B

Chapter 2

Physical Geology

eighth edition

Plummer/McGeary/Carlson

MINERALS

MINERAL vs. ROCK

- ROCK
  - An aggregate of MINERALS (usually)
- MINERAL
  - Crystalline- orderly arrangement of atoms
  - Naturally occurring
  - Inorganic
  - Definite chemical composition
    - e.g. SiO_2 for quartz; $KAlSi_3O_8$ for feldspar

Atom & Elements

- Atoms
  - Nucleus
    - Proton, neutron
  - Electron
- Molecule- e.g. water molecule

Element

- Atomic number
- Isotope
- Atomic weight

Chemical activity

- Stable atoms want
  - positive & negative charges balanced
  - electron shells full
- Ions
- Bonding
  - Ionic
  - Covalent
  - Metallic
  - Van der Waal's

Chemical composition of the Crust

- Oxygen most abundant
- Followed by silicon and aluminum
- The most common minerals will be composed mostly of these elements
- Silica & silicates

Crystallinity

- 3 dimensional orderliness of atoms
- Silicon-oxygen tetrahedron
- Silicate structures
  - Isolated- e.g. olivine
  - Chain
    - Single chain- pyroxene
    - Double chain- amphibole
  - Sheet- e.g. mica, clay
  - Framework- e.g. quartz, feldspar

MINERALS

- Crystalline solids
- Natural and Inorganic Substances
- Definite chemical composition
  - Can be written as a chemical formula
  - Solid solution
    - Zoning

Important Minerals

- Quartz
- FELDSPAR Group
  - *Potassium Feldspar*
  - *Plagioclase Feldspar*
    - Sodium (Na) , Calcium (Ca) Feldspar

Important Minerals

- PYROXENE Group- Augite most common
- AMPHIBOLE Group- Hornblende most common
- MICA Group- Si + O in sheets
  - Biotite
  - Muscovite
- CALCITE- $CaCO_3$

Properties of Minerals

- Color
  - Not always reliable
  - *Ferromagnesian* minerals green or black
- Streak
- Luster
  - Metallic
  - Nonmetallic
    - Vitreous or Glassyl
    - Earthy

Properties of Minerals

- Hardness
- Moh's Hardness Scale
  - Fingernail = 2.5
  - Penny = 3.5
  - Knife/Glass = 5.5
- Crystal Form
  - Law of Constancy of Interfacial Angles

Properties of Minerals

- **Cleavage**
  - Quality
  - Number of directions
    - One- e.g. *Mica*
    - Two at right angles- e.g. *Feldspar, Pyroxene*
    - Two not at right angles- e.g. *Amphibole*
    - Three at right angles (cubic)- e.g. *Halite*
    - Three not at right angles (rhombohedral)- e.g. calcite
    - Four or six- not common

Properties of Minerals

- Fracture
  - Absence of cleavage
  - Irregular fracture
  - Conchoidal fracture
- Density
  - Specific Gravity
- Some unusual properties
  - Striations, Magnetism, Taste, Double refraction

Chemical tests

- Reaction with HCl
  - Calcite effervesces

ROCK CYCLE

- Equilibrium
- Interrelationships between
  - igneous rocks
  - sediment
  - sedimentary rocks
  - metamorphic rocks
  - weathering and erosion
- Plate Tectonic Example

Specimen of Granite

Feldspar
Quartz
Biotite
5 mm
0.0000001 mm
Silicon and oxygen atoms in crystalline structure
Diagrammatic representation of crystalline structure

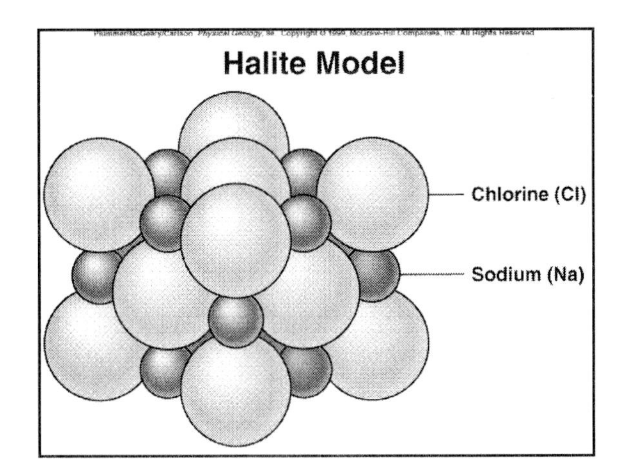
Halite Model
Chlorine (Cl)
Sodium (Na)

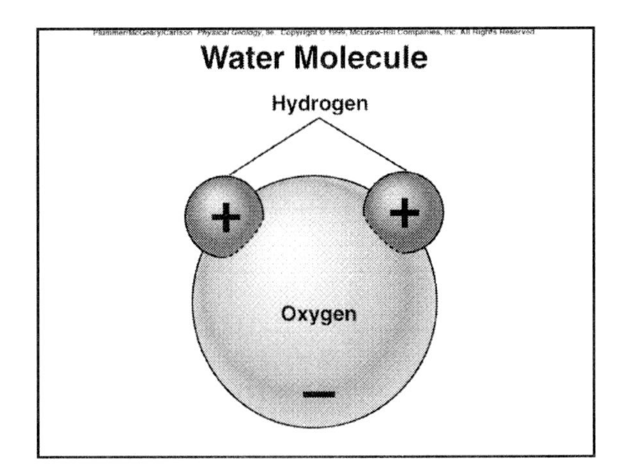
Water Molecule
Hydrogen
+
+
Oxygen
–

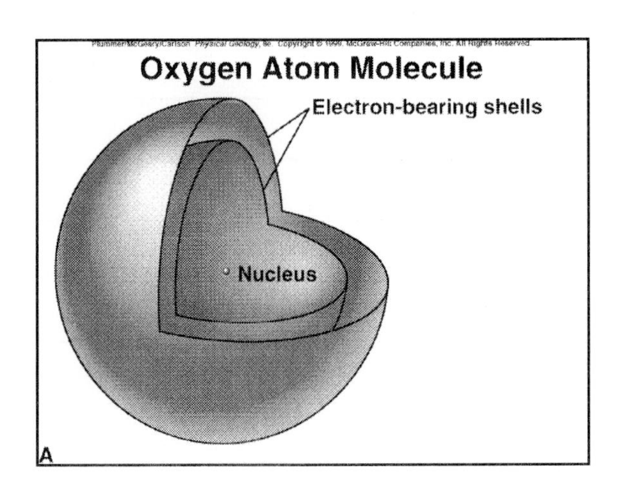
Oxygen Atom Molecule
Electron-bearing shells
Nucleus
A

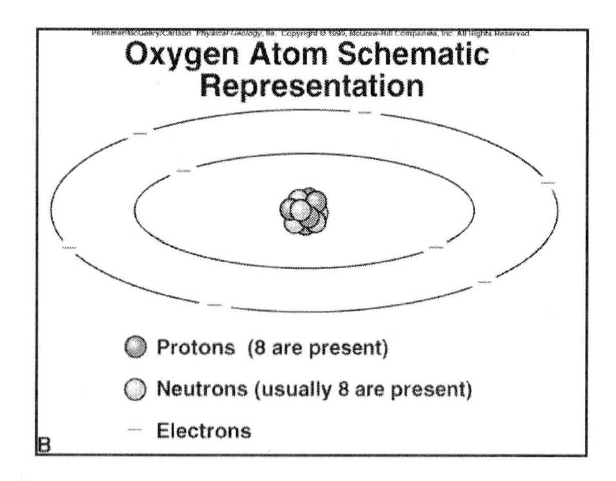
Oxygen Atom Schematic Representation
Protons (8 are present)
Neutrons (usually 8 are present)
Electrons
B

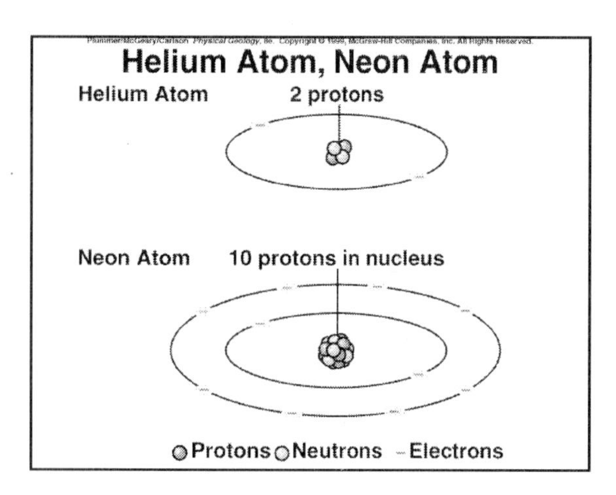
Helium Atom, Neon Atom
Helium Atom
2 protons
Neon Atom
10 protons in nucleus
Protons
Neutrons
Electrons

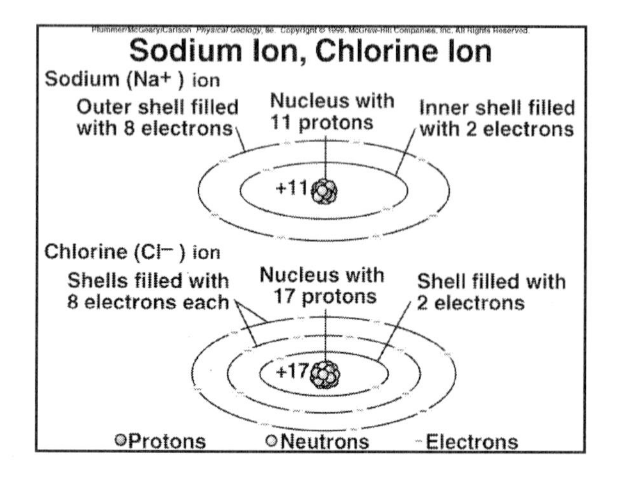
Sodium Ion, Chlorine Ion
Sodium (Na+) ion
Outer shell filled with 8 electrons
Nucleus with 11 protons
Inner shell filled with 2 electrons
+11
Chlorine (Cl−) ion
Shells filled with 8 electrons each
Nucleus with 17 protons
Shell filled with 2 electrons
+17
Protons
Neutrons
Electrons

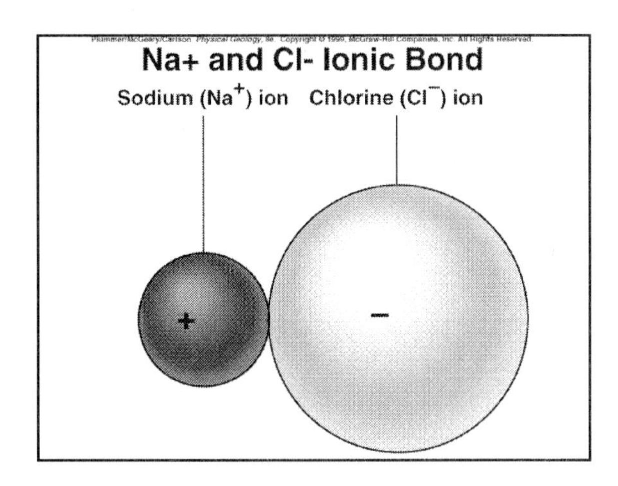
Na+ and Cl- Ionic Bond
Sodium (Na+) ion
Chlorine (Cl−) ion
+
−

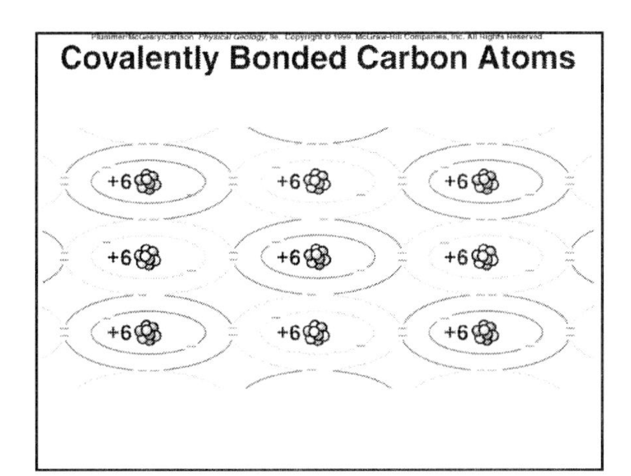
Covalently Bonded Carbon Atoms
+6
+6
+6
+6
+6
+6
+6
+6
+6

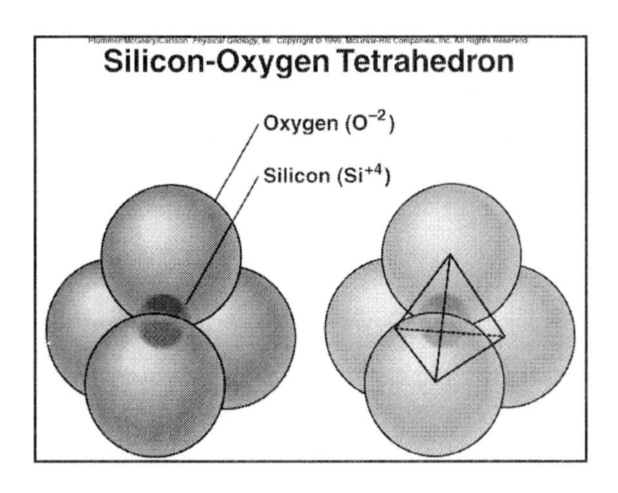
Silicon-Oxygen Tetrahedron
Oxygen (O−2)
Silicon (Si+4)

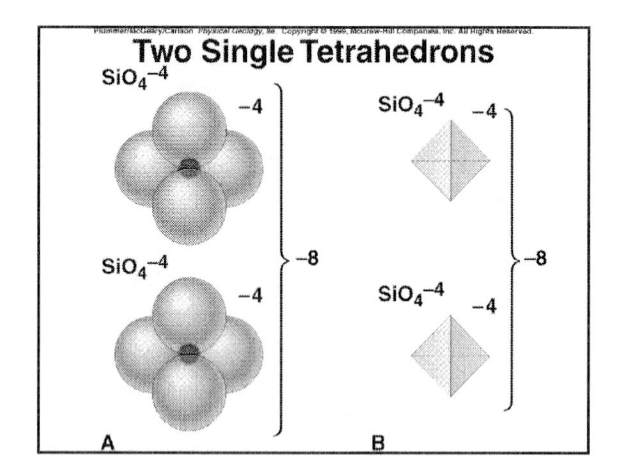
Plummer/McGeary/Carlson Physical Geology, 8e. Copyright © 1999, McGraw-Hill Companies, Inc. All Rights Reserved.
Two Single Tetrahedrons
SiO_4^{-4}
−4
SiO_4^{-4}
−4
−8
SiO_4^{-4}
−4
SiO_4^{-4}
−4
−8
A
B

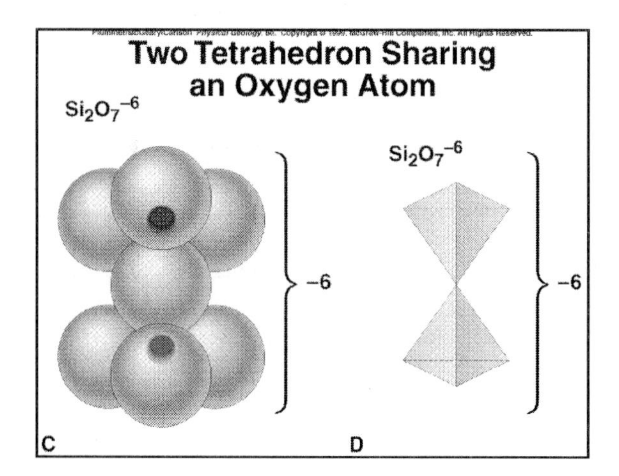
Two Tetrahedron Sharing an Oxygen Atom
$Si_2O_7^{-6}$
$Si_2O_7^{-6}$
−6
−6
C
D

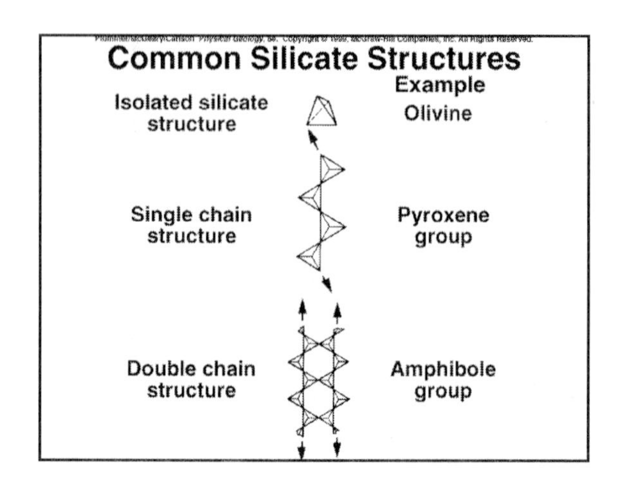
Common Silicate Structures
Example
Isolated silicate structure
Olivine
Single chain structure
Pyroxene group
Double chain structure
Amphibole group

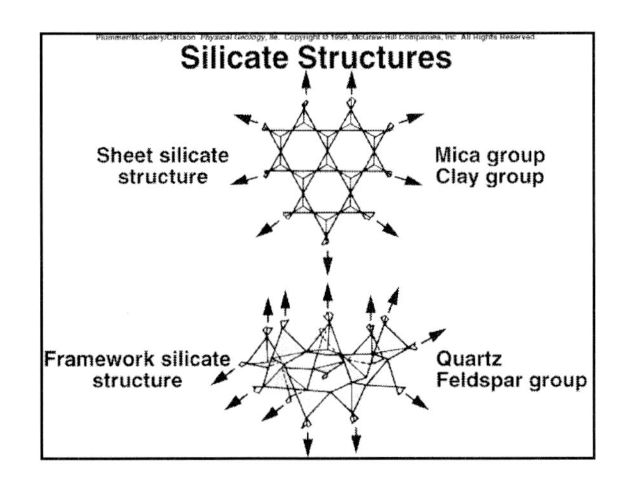

Silicate Structures
Sheet silicate
structure
Mica group
Clay group
Framework silicate
structure
Quartz
Feldspar group

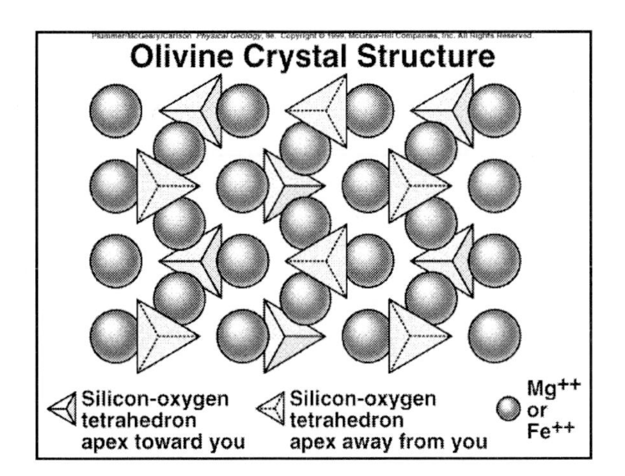

Olivine Crystal Structure
Silicon-oxygen
tetrahedron
apex toward you
Silicon-oxygen
tetrahedron
apex away from you
Mg++
or
Fe++

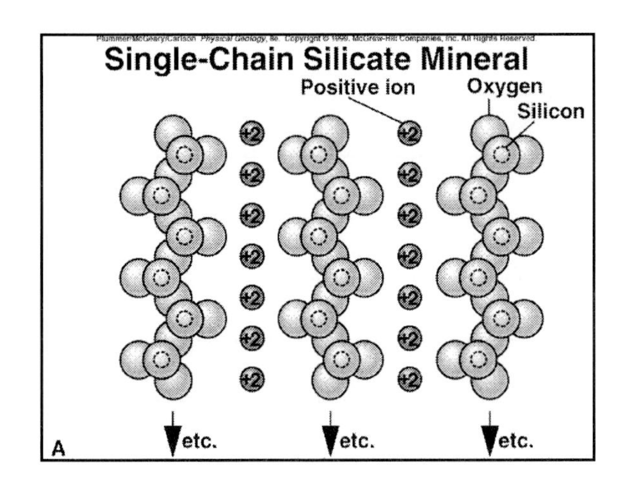

Single-Chain Silicate Mineral
Positive ion
Oxygen
Silicon
+2
etc.
etc.
etc.
A

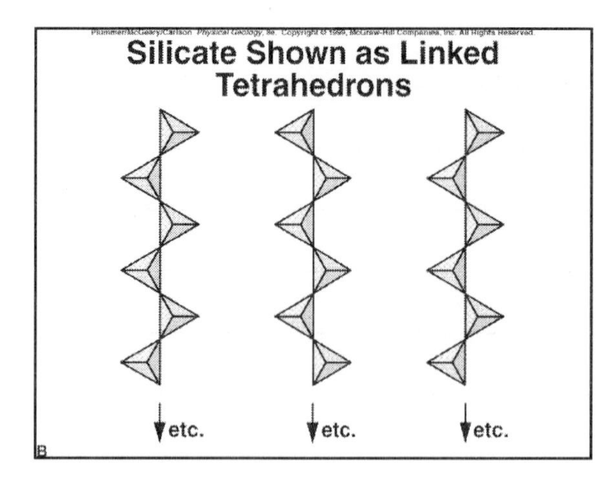
Silicate Shown as Linked Tetrahedrons
etc.
etc.
etc.
B

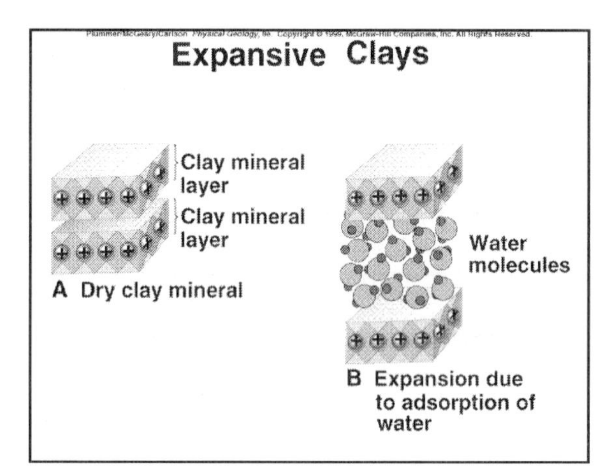
Expansive Clays
Clay mineral layer
Clay mineral layer
A Dry clay mineral
Water molecules
B Expansion due to adsorption of water

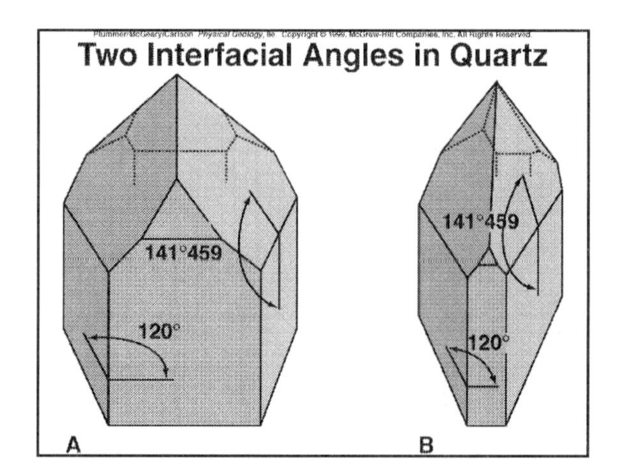
Two Interfacial Angles in Quartz
141°459
120°
141°459
120°
A
B

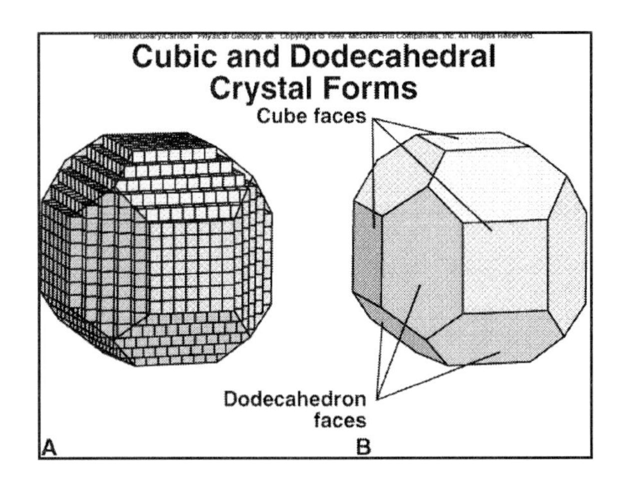
Cubic and Dodecahedral Crystal Forms
Cube faces
Dodecahedron faces
A
B

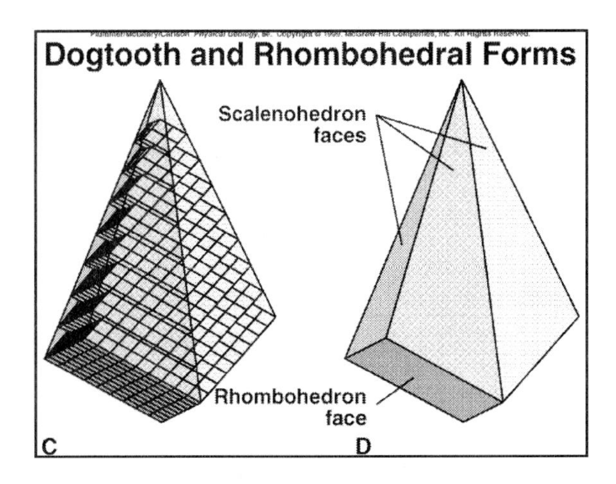
Dogtooth and Rhombohedral Forms
Scalenohedron faces
Rhombohedron face
C
D

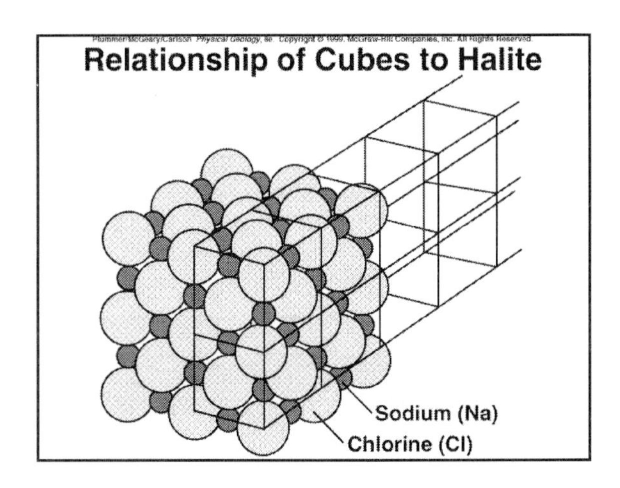
Relationship of Cubes to Halite
Sodium (Na)
Chlorine (Cl)

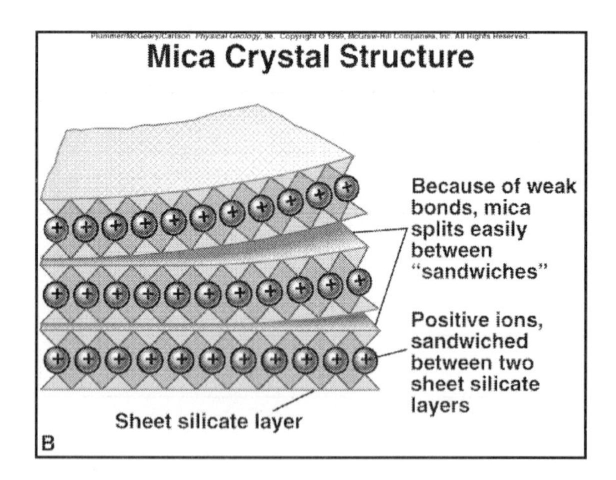
Mica Crystal Structure
Because of weak bonds, mica splits easily between "sandwiches"
Positive ions, sandwiched between two sheet silicate layers
Sheet silicate layer
B

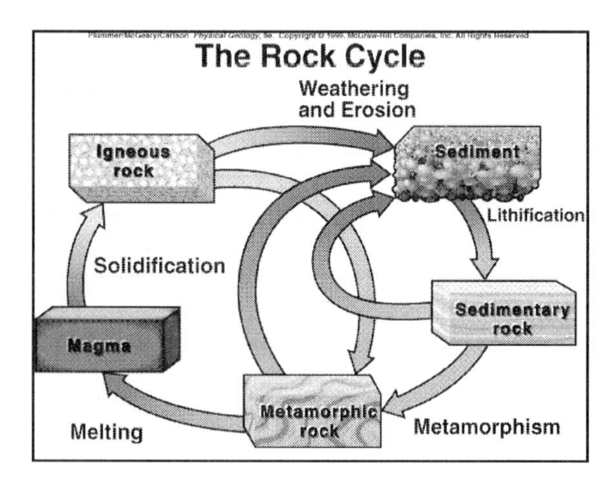
The Rock Cycle
Weathering and Erosion
Igneous rock
Sediment
Lithification
Solidification
Sedimentary rock
Magma
Metamorphic rock
Melting
Metamorphism

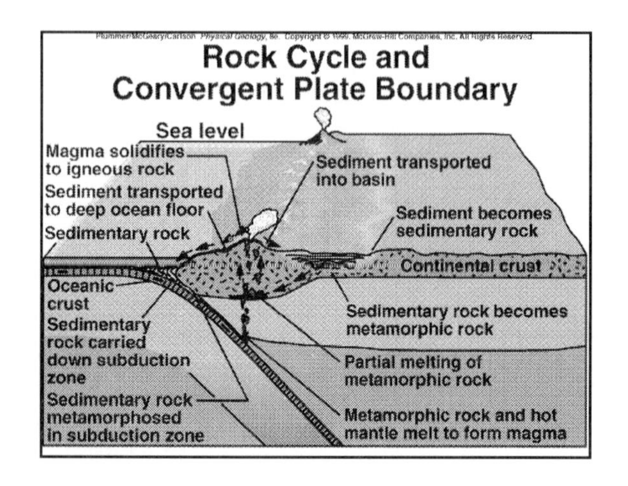
Rock Cycle and Convergent Plate Boundary
Sea level
Magma solidifies to igneous rock
Sediment transported to deep ocean floor
Sedimentary rock
Sediment transported into basin
Sediment becomes sedimentary rock
Continental crust
Oceanic crust
Sedimentary rock carried down subduction zone
Sedimentary rock metamorphosed in subduction zone
Sedimentary rock becomes metamorphic rock
Partial melting of metamorphic rock
Metamorphic rock and hot mantle melt to form magma

Chapter 3

Physical Geology

eighth edition

Plummer/McGeary/Carlson

IGNEOUS ROCKS

EXTRUSIVE (Volcanic)- Fine-grained

INTRUSIVE (Plutonic)- Coarse-grained

MAGMA

- Molten Rock
  - Usually with dissolved gasses
- Generated at depth
- Eruptions if magma (lava) reaches surface
- If doesn't reach surface, Solidifies underground
  - Intrudes *country rock*
  - Intrusive *contact*
  - *Chill zone*
  - *Zenolith*

Igneous Rocks

- Names based on ***mineral*** composition (which reflects chemical composition of the magma) and... Grain size
  - Course-grained: > 1 mm.
  - Fine-grained: < 1 mm.
  - Porphyritic

Igneous Rocks-Classification

- **Course-grained-Plutonic**
- GRANITE
- Diorite
- Gabbro
- **Fine-Grained**
- Rhyolite
- ANDESITE*
- BASALT*

Igneous Rock Identification

- Granite (& Rhyolite)
  - High in Si + O
  - Low in Fe + Mg
  - Mostly feldspar & quartz
  - Light-colored
- Basalt (& Gabbro)
  - "Low" in Si + O
  - High in Fe + Mg
  - no quartz, abundant ferromagnesian minerals
  - Dark colored
- Andesite (& Diorite- intermediate)

Chemistry of Igneous Rocks

- Mafic rocks or magma
- Silicic (or felsic) rocks or magma
- Intermediate rocks or magma
- Ultramafic rocks

INTRUSIVE BODIES (STRUCTURES)

- Bodies that solidified underground
- Volcanic neck- shallow intrusion
- Fills cracks- tabular bodies
  - DIKE-
    - If no layering in ***country rock***
    - If country rock is layered- Discordant
  - SILL- less common
    - Concordant- parallel to layering in country rock

INTRUSIVE STRUCTURES

- Pluton
- BATHOLITH-
  - Large intrusive body
  - Exposed in an area greater than 100 square Km.
  - Coalesced smaller plutons
- smaller bodies are called STOCKS
- Batholiths a gathering of smaller blobs
- Magma moves upward from depth as *diapirs*

DISTRIBUTION OF PLUTONIC ROCK

- Granite most abundant
  - Common in mountain ranges
  - In ancient rock that were mountain ranges that are now plains
- Ultramafic rock in the mantle

How magma forms

- Partial melting of rock at depth
- Source of heat
  - Geothermal gradient
  - Mantle plumes
- Factors that control melting temperatures
  - Pressure
  - Water under pressure
  - Mixing of minerals

How magmas evolve

- Differentiation
  - Bowen's Reaction Series
    - Discontinuous Branch
    - Continuous Branch
  - Crystal Settling
    - Ore deposits due to crystal settling
- Partial melting
- Assimilation
- Mixing of magmas

PLATE TECTONICS & IGNEOUS ACTIVITY

- DIVERGENT BOUNDARY
  - Notably at mid-oceanic ridges
  - Sea floor Spreading
  - Magma from asthenosphere
    - Partial melting
      - Due to reduced pressure
    - Produces mafic magma
    - Solidifies into basalt and gabbro
    - Becomes oceanic crust
    - Unmelted residue remains as ultramafic rock

PLATE TECTONICS

- INTRAPLATE IGNEOUS ACTIVITY
  - Attributed to mantle plumes

PLATE TECTONICS

- CONVERGENT BOUNDARY
  - Origin of Andesite
    - Partial melting of asthenosphere above subducted crust
    - Water lowers melting temperature producing mafic magma
    - Ascending magma modified into intermediate magma
  - Origin of Granite
    - Partial melting of lower continental crust
    - Heat from mafic magma *underplating* the crust

PLATE TECTONICS

- Alternate hypotheses for explaining andesitic and granitic magmas
  - Partial melting of basalt in subduction zone
  - Assimilation of crustal rocks
  - Mixing of magmas
  - Melting of sedimentary rocks

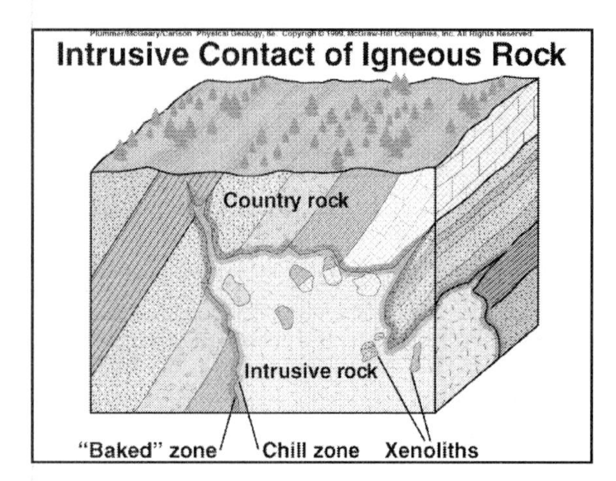

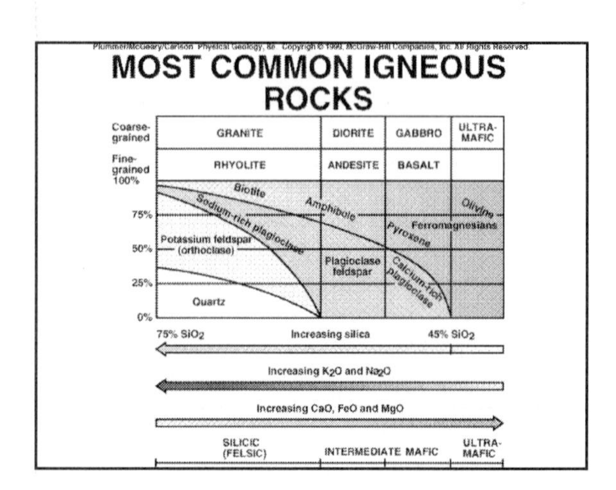

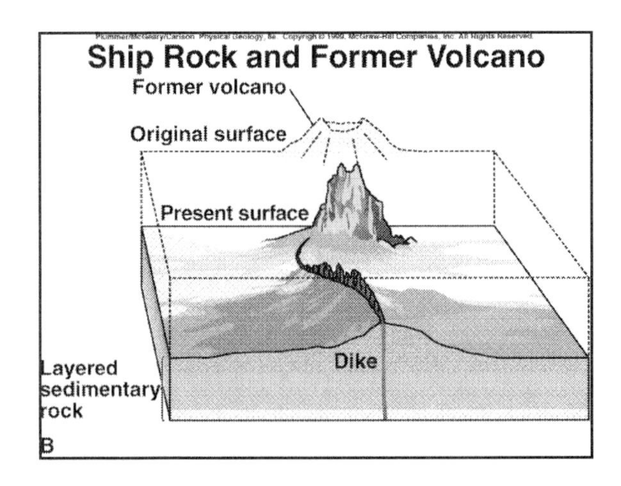
Ship Rock and Former Volcano
Former volcano
Original surface
Present surface
Dike
Layered sedimentary rock
B

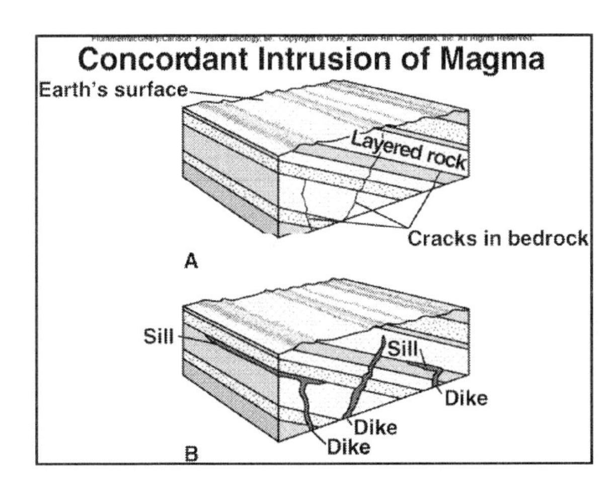
Concordant Intrusion of Magma
Earth's surface
Layered rock
Cracks in bedrock
A
Sill
Sill
Dike
Dike
Dike
B

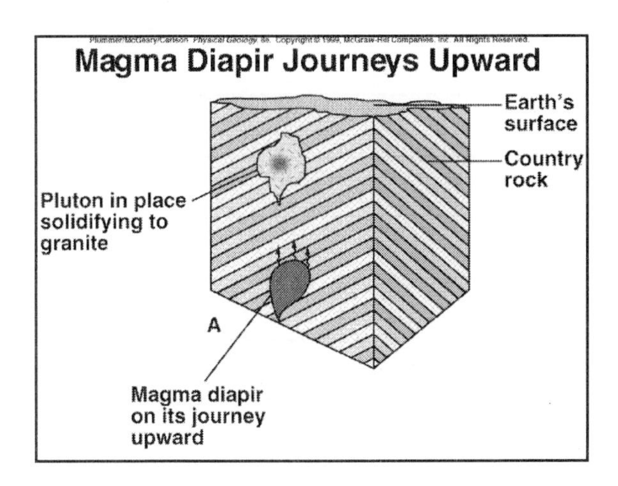
Magma Diapir Journeys Upward
Earth's surface
Country rock
Pluton in place solidifying to granite
A
Magma diapir on its journey upward

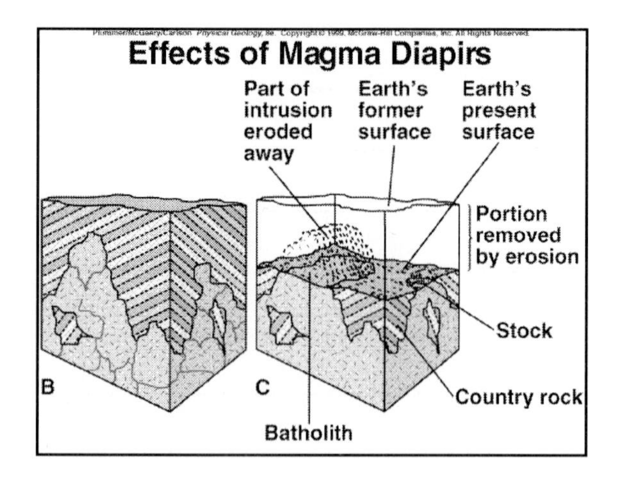
Plummer/McGeary/Carlson Physical Geology, 8e. Copyright © 1999, McGraw-Hill Companies, Inc. All Rights Reserved.
Effects of Magma Diapirs
Part of intrusion eroded away
Earth's former surface
Earth's present surface
Portion removed by erosion
Stock
Country rock
Batholith
B
C

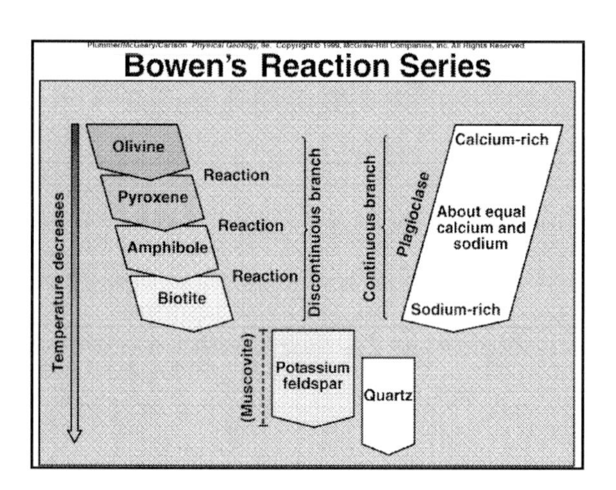
Plummer/McGeary/Carlson Physical Geology, 8e. Copyright © 1999, McGraw-Hill Companies, Inc. All Rights Reserved.
Bowen's Reaction Series
Temperature decreases
Olivine
Reaction
Pyroxene
Reaction
Amphibole
Reaction
Biotite
Discontinuous branch
Continuous branch
Plagioclase
Calcium-rich
About equal calcium and sodium
Sodium-rich
(Muscovite)
Potassium feldspar
Quartz

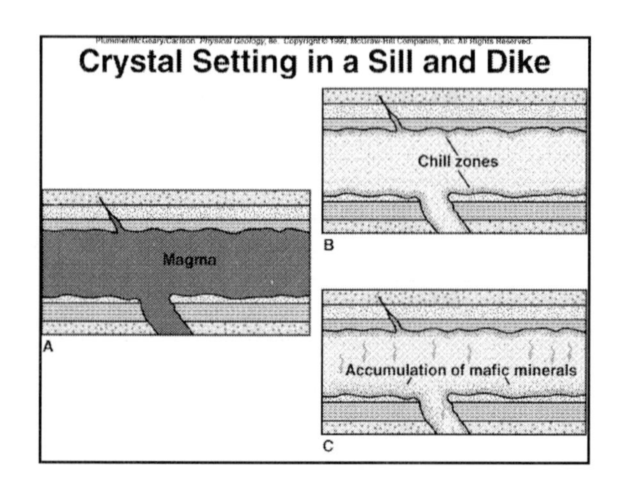
Plummer/McGeary/Carlson Physical Geology, 8e. Copyright © 1999, McGraw-Hill Companies, Inc. All Rights Reserved.
Crystal Setting in a Sill and Dike
Magma
A
Chill zones
B
Accumulation of mafic minerals
C

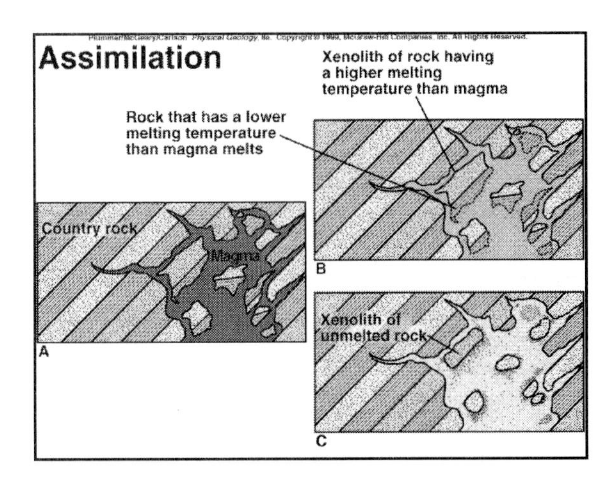
Assimilation
Xenolith of rock having
a higher melting
temperature than magma
Rock that has a lower
melting temperature
than magma melts
Country rock
Magma
A
B
Xenolith of
unmelted rock
C

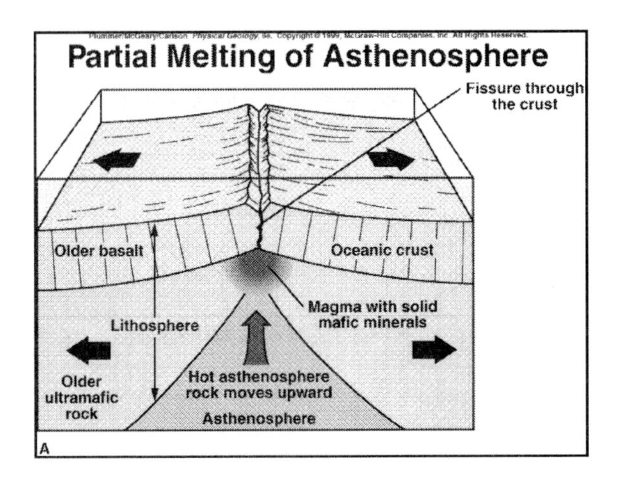
Partial Melting of Asthenosphere
Fissure through
the crust
Older basalt
Oceanic crust
Magma with solid
mafic minerals
Lithosphere
Older
ultramafic
rock
Hot asthenosphere
rock moves upward
Asthenosphere
A

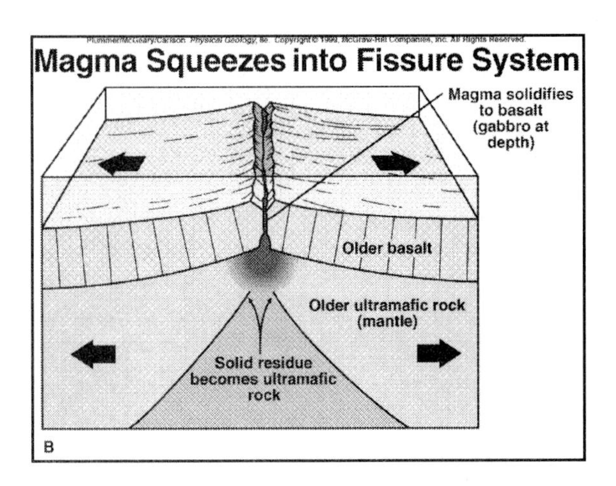
Magma Squeezes into Fissure System
Magma solidifies
to basalt
(gabbro at
depth)
Older basalt
Older ultramafic rock
(mantle)
Solid residue
becomes ultramafic
rock
B

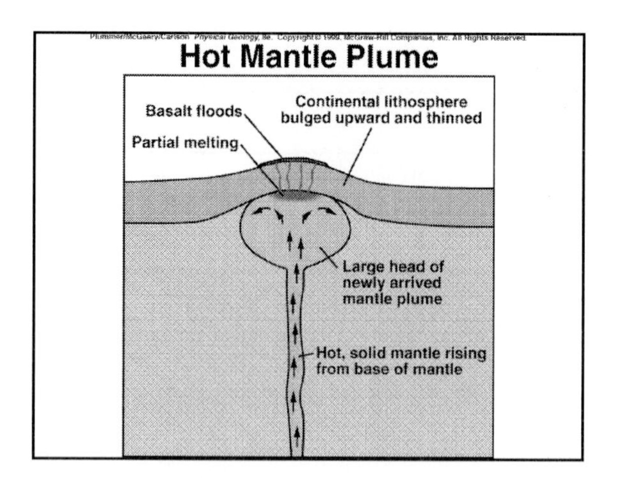
Hot Mantle Plume
Basalt floods
Continental lithosphere bulged upward and thinned
Partial melting
Large head of newly arrived mantle plume
Hot, solid mantle rising from base of mantle

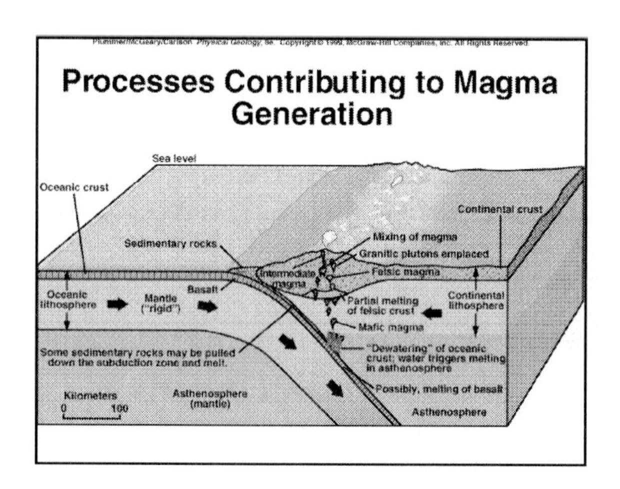
Processes Contributing to Magma Generation
Sea level
Oceanic crust
Continental crust
Sedimentary rocks
Mixing of magma
Granitic plutons emplaced
Intermediate magma
Felsic magma
Basalt
Oceanic lithosphere
Mantle ("rigid")
Partial melting of felsic crust
Continental lithosphere
Mafic magma
"Dewatering" of oceanic crust; water triggers melting in asthenosphere
Some sedimentary rocks may be pulled down the subduction zone and melt.
Possibly, melting of basalt
Kilometers
0 100
Asthenosphere (mantle)
Asthenosphere

Chapter 4

Physical Geology

eighth edition

Plummer/McGeary/Carlson

VOLCANISM & EXTRUSIVE ROCKS

VOLCANISM

- Lava = Magma at earth surface
  - Silica content controls "explosiveness"
- Pyroclasts = Fragments of rock due to explosion
- Lava flows
- Extrusive rocks
- Volcano

Effects on Humans

- Growth of Hawaii
- Geothermal energy
- Effect on climate
- Volcanic catastrophies
  - Mt. St. Helens
  - Vesuvius
  - Krakatoa
  - Crater Lake

Eruptive Violence & Characteristics of Lava

- Gas in lava
- Viscosity
  - Temperature
  - Silica content
    - Silicic lavas- most viscous
    - Mafic lavas- least viscous

Extrusive Rocks & Gases

- Scientific study of volcanism
- Gases
  - Primarily H_2O
  - Also CO_2 , SO_2 , H_2S, HCl
- Gases & pyroclastics
  - Ashfall
  - Pyroclastic flow

Extrusive Rocks

- Importance of silica content
- Rhyolite- silicic
  - Predominantly feldspar and quartz
- Andesite- intermediate
  - Plagioclase feldspar & ferromagnesian minerals
- Basalt- mafic
  - Ferromagnesian minerals & plagioclase feldspar

Extrusive Rocks

- Textures
  - Fine-grained (smaller than 1 mm)
  - Glassy- *Obsidian*
  - Due to
    - rapid cooling (mainly)
    - high viscosity
  - Porphyritic
    - Phenocrysts
  - Due to trapped gas
    - Vesicles

Extrusive Rocks

- Textures
  - Fine-grained (smaller than 1 mm)
  - Glassy- *Obsidian*
  - Due to
    - rapid cooling (mainly)
    - high viscosity
  - Porphyritic
    - Phenocrysts

Extrusive Rocks

- Textures
  - Due to trapped gas
    - Vesicles
    - Scoria
    - Pumice
  - Fragmental
    - Pyroclasts
    - Dust, ash, cinders
    - Blocks & bombs
    - Tuff
    - Volcanic Breccia

VOLCANOES

- Volcanoes are cone-shaped
- Vent
- Crater
- Flank eruption
- Caldera
- Types:
  - Shield, Cinder Cone, Composite

SHIELD VOLCANOES

- Low viscosity lava flows
  - Low silica magma- mafic
  - Basalt
    - *Pahoehoe*
    - *Aa*
- Gently sloping flanks- between 2 and 10 degrees
- Tend to be very large
- Spatter Cone- minor feature

CINDER CONES

- Formed of pyroclastics only
- Steep sides- ~30 degrees
- Relatively small
- Short duration of activity

COMPOSITE VOLCANO

- Alternating pyroclastic layers & lava flows
- Slopes intermediate in steepness
- Intermittent eruptions over long time span
- Mostly *Andesite*
- *Distribution*
  - Circum-Pacific Belt ("Ring of Fire")
  - Mediterranean Belt

VOLCANIC DOMES

- Forms above a volcanic vent
- Viscous lava
  - Usually silica-rich (or cooler magma)
- Associated with violent eruptions

LAVA FLOODS

- Mafic lava- solidifies to basalt
- Fissure flows
  - Plateau basalts
- Columnar structure or jointing

SUBMARINE ERUPTIONS

- Pillow basalt

Events of 1980 Eruption of Mt. St. Helens

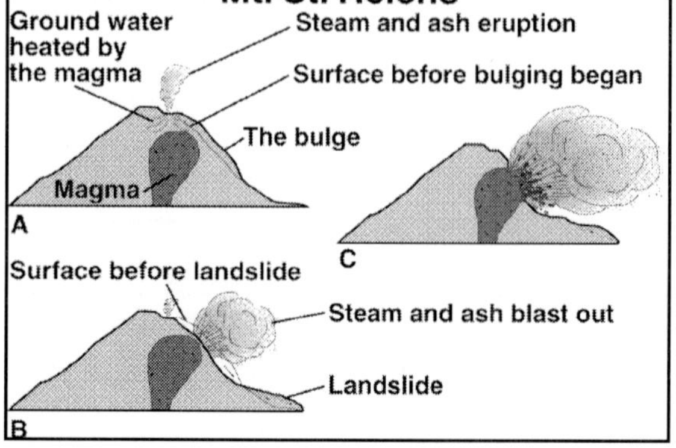

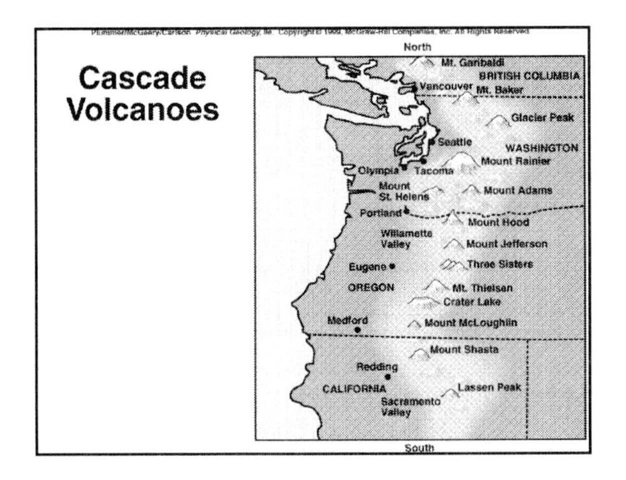
Plummer/McGeary/Carlson Physical Geology, 8e Copyright © 1999, McGraw-Hill Companies, Inc. All Rights Reserved.
Cascade Volcanoes
North
Mt. Garibaldi
BRITISH COLUMBIA
Vancouver
Mt. Baker
Glacier Peak
Seattle
WASHINGTON
Mount Rainier
Olympia
Tacoma
Mount St. Helens
Mount Adams
Portland
Mount Hood
Willamette Valley
Mount Jefferson
Eugene
Three Sisters
OREGON
Mt. Thielsen
Crater Lake
Medford
Mount McLoughlin
Mount Shasta
Redding
CALIFORNIA
Lassen Peak
Sacramento Valley
South

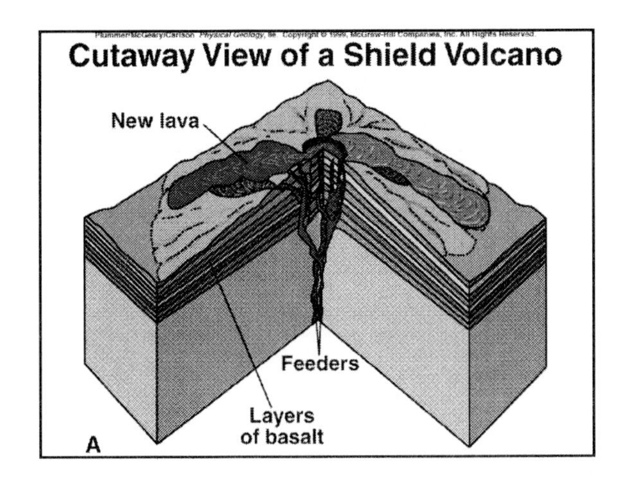
Plummer/McGeary/Carlson Physical Geology, 8e Copyright © 1999, McGraw-Hill Companies, Inc. All Rights Reserved.
Cutaway View of a Shield Volcano
New lava
Feeders
Layers of basalt
A

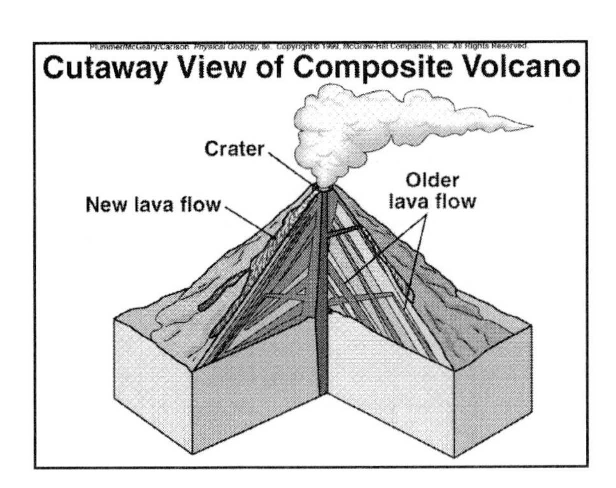
Plummer/McGeary/Carlson Physical Geology, 8e Copyright © 1999, McGraw-Hill Companies, Inc. All Rights Reserved.
Cutaway View of Composite Volcano
Crater
New lava flow
Older lava flow

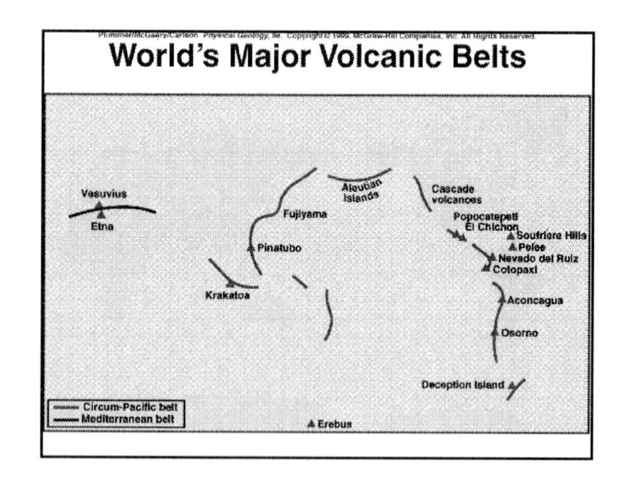
Plummer/McGeary/Carlson Physical Geology, 8e. Copyright © 1999, McGraw-Hill Companies, Inc. All Rights Reserved.
World's Major Volcanic Belts
Vesuvius
Etna
Fujiyama
Aleutian Islands
Cascade volcanoes
Popocatepetl
El Chichon
Soufriere Hills
Pelee
Nevado del Ruiz
Cotopaxi
Pinatubo
Krakatoa
Aconcagua
Osorno
Deception Island
Circum-Pacific belt
Mediterranean belt
Erebus

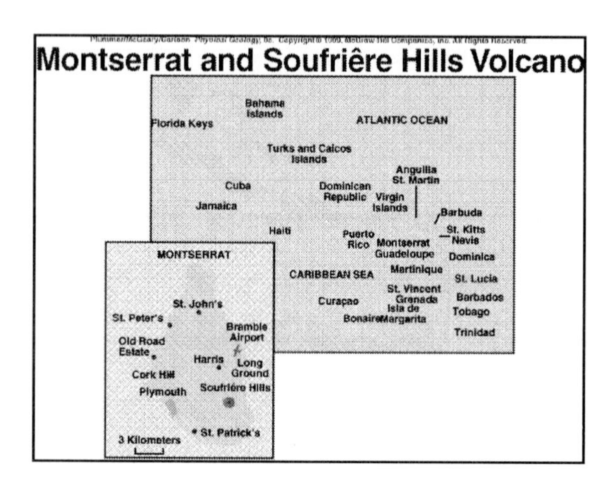
Montserrat and Soufrière Hills Volcano
Bahama Islands
Florida Keys
ATLANTIC OCEAN
Turks and Caicos Islands
Anguilla
St. Martin
Cuba
Dominican Republic
Virgin Islands
Jamaica
Barbuda
Haiti
Puerto Rico
St. Kitts
Nevis
Montserrat
Guadeloupe
Dominica
MONTSERRAT
CARIBBEAN SEA
Martinique
St. Lucia
St. Vincent
Barbados
Curaçao
Grenada
Isla de Margarita
Bonaire
Tobago
Trinidad
St. John's
St. Peter's
Bramble Airport
Old Road Estate
Harris
Long Ground
Cork Hill
Plymouth
Soufrière Hills
St. Patrick's
3 Kilometers

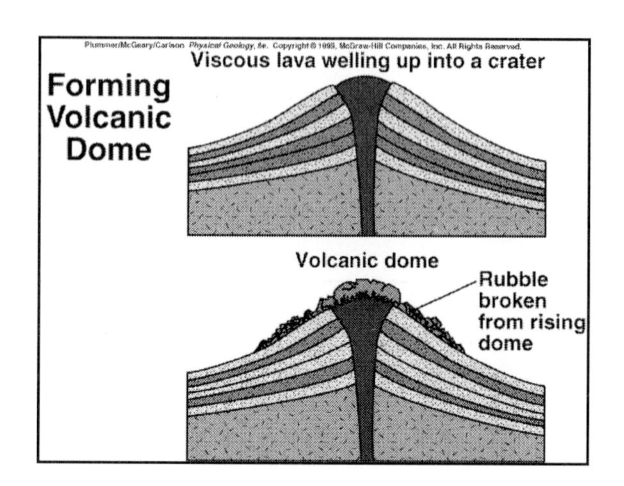
Plummer/McGeary/Carlson Physical Geology, 8e. Copyright © 1999, McGraw-Hill Companies, Inc. All Rights Reserved.
Forming Volcanic Dome
Viscous lava welling up into a crater
Volcanic dome
Rubble broken from rising dome

Chapter 5

Physical Geology

eighth edition

Plummer/McGeary/Carlson

WEATHERING & SOIL

WEATHERING, EROSION, TRANSPORTATION

- *Weathering*- Rock breaks down
- *Erosion*- Physical removal
- *Transportation*- Movement of eroded particles
- Chemical vs. Physical Weathering
- Effects of weathering
  - Surface alteration of outcrops
  - Spheroidal weathering
  - Differential weathering

MECHANICAL WEATHERING

- Disintegration of rocks
- Frost Action
  - Frost wedging - rock pried apart along *joints*
    - water expands upon freezing
  - Frost heaving
- Abrasion during transportation

MECHANICAL WEATHERING

- Pressure Release
  - sheet joint
  - exfoliation
    - exfoliation domes
- Other
  - Plant growth; Burrowing animals
  - Pressure of salt crystals
  - Extreme temperature changes
- Mechanical weathering increases surface area facilitating chemical weathering

CHEMICAL WEATHERING

- Decomposition of rock to form new substances
  - Equilibrium
- Role of Oxygen
  - Fe in ferromagnesian minerals becomes oxidized
    - Hematite
    - Limonite

Chemical Weathering

- Role of Acid
  - Hydrogen ions in water
- Carbonic Acid H_2CO_3
  - $CO_2 + H_2O \rightarrow H_2CO_3$
  - $H_2CO_3 \rightarrow H^+ + HCO_3^-$ (in water)
  - Solution weathering
    - Calcite (in limestone) dissolves in acidic water
    - fluted (channeled) rock surface
    - caves

Chemical Weathering

- Chemical weathering of feldspar
  - H^+ + feldspar -> clay mineral
- Chemical weathering of other minerals
- Weathering & diamond concentration
- Weathering and climate
- Weathering Products

SOIL

- Various definitions
  - Unconsolidated material above bedrock
  - weathered material & organic matter
    - supports plant life
- Loam
- *Clay-sized particles* vs. *clay minerals*
- Soil Horizons
- Residual Soil
- Transported Soil

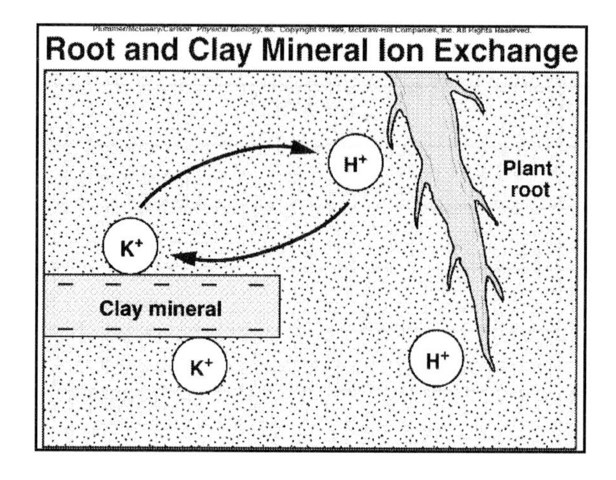
Root and Clay Mineral Ion Exchange
H+
Plant root
K+
Clay mineral
K+
H+

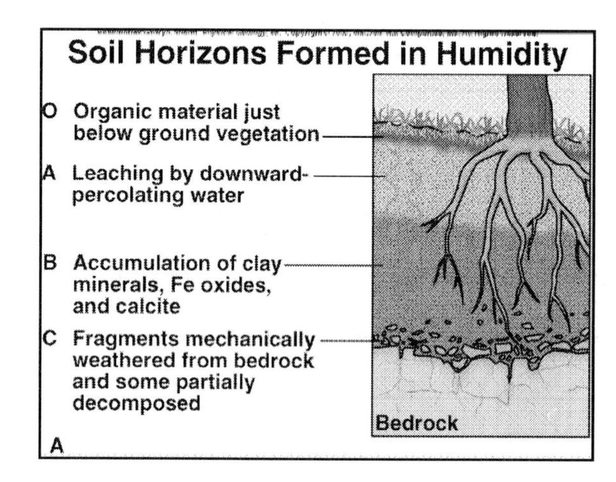
Soil Horizons Formed in Humidity
O Organic material just below ground vegetation
A Leaching by downward-percolating water
B Accumulation of clay minerals, Fe oxides, and calcite
C Fragments mechanically weathered from bedrock and some partially decomposed
Bedrock
A

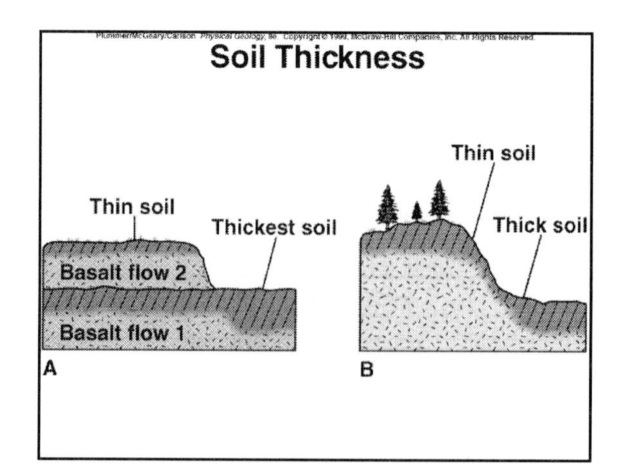
Soil Thickness
Thin soil
Thickest soil
Basalt flow 2
Basalt flow 1
A
Thin soil
Thick soil
B

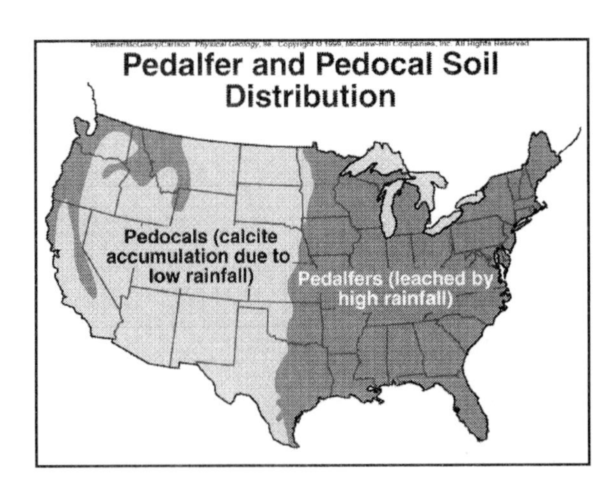
Pedalfer and Pedocal Soil Distribution
Pedocals (calcite accumulation due to low rainfall)
Pedalfers (leached by high rainfall)

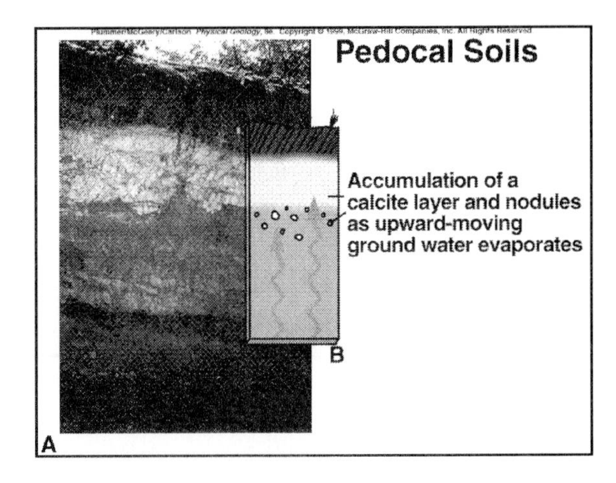
Pedocal Soils
Accumulation of a calcite layer and nodules as upward-moving ground water evaporates
A
B

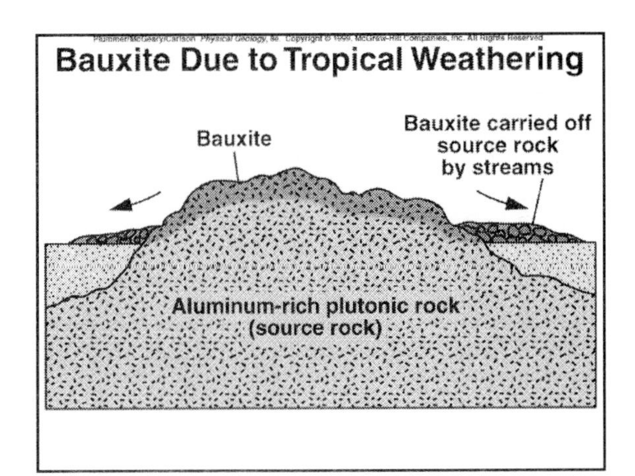
Bauxite Due to Tropical Weathering
Bauxite
Bauxite carried off source rock by streams
Aluminum-rich plutonic rock (source rock)

Chapter 6

Physical Geology

eighth edition

Plummer/McGeary/Carlson

SEDIMENTS & SEDIMENTARY ROCKS

SEDIMENT

- From
  - Weathering & Erosion
  - Precipitation from solution
- Unconsolidated

SEDIMENT

- Particle size
  - Gravel- Pebbles, cobbles, boulders
  - Sand
  - Silt
  - Clay
    - Deposition*Clay-sized* particle vs. *clay mineral*

SEDIMENT

- Transportation
  - Rounding
  - Sorting
  - Deposition
    - Environment of deposition
- Preservation

SEDIMENT

- Lithification
  - Compaction
    - Decrease in *pore space*
  - Cementation
    - Clastic texture
  - Crystallization
    - Crystalline texture

TYPES OF SEDIMENTARY ROCKS

- Clastic, Chemical, Organic

CLASTIC ROCKS

- Breccia and Conglomerate
  - Sedimentary Breccia- angular fragments
  - Conglomerate- rounded fragments
- Sandstone
  - Quartz sandstone
  - Arkose
  - Graywacke
    - Fine-grained *Matrix*
    - Usually from *turbidity currents*

CLASTIC ROCKS

- Fine-grained Rocks
  - Shale
  - Siltstone
  - Claystone
  - Mudstone

CHEMICAL SEDIMENTARY ROCKS

- Carbonate Rocks
  - Limestone- made of calcite
    - Biochemical varieties
    - Inorganic varieties
  - Dolomite
  - Recrystallization
- Chert
- Evaporites
  - Rock gypsum
  - Rock salt

ORGANIC SEDIMENTARY ROCKS

- Coal
  - Develops from peat

ORIGIN OF OIL AND GAS

- Formed from microscopic organism
- Partially decomposed
- Buried
- Convert to oil and gas at higher temperatures
- Move upward into sandy layers

SEDIMENTARY STRUCTURES

- BEDDING-
  - Principle of Original Horizontality
  - Bedding plane
  - Cross-bedding
  - Graded bed
  - Mud cracks
  - Ripple marks
  - Fossils

FORMATIONS

- Large enough to show on a map
- Distinctive from neighboring rock units
- Named after geographic locations
- CONTACT
  - Sedimentary contact
  - Other contacts

INTERPRETATION OF SEDIMENTARY ROCKS

- Source Area
- Environment of Deposition
  - Glacial environments
  - Alluvial fan
  - River channel and flood plain
  - Lake
  - Delta

INTERPRETATION OF SEDIMENTARY ROCK

- Environment of Deposition
  - Beach, Barrier Island, Dune
  - Lagoon
  - Shallow marine shelf
  - Reef
  - Deep marine environment

INTERPRETATION OF SEDIMENTARY ROCK

- Plate Tectonics and Sedimentary Rocks
  - At convergent boundaries
  - At transform boundaries
  - At divergent boundaries

Sediment Sorting By a River

Mountains

River

Plain

Gravel

Sand

Silt and clay

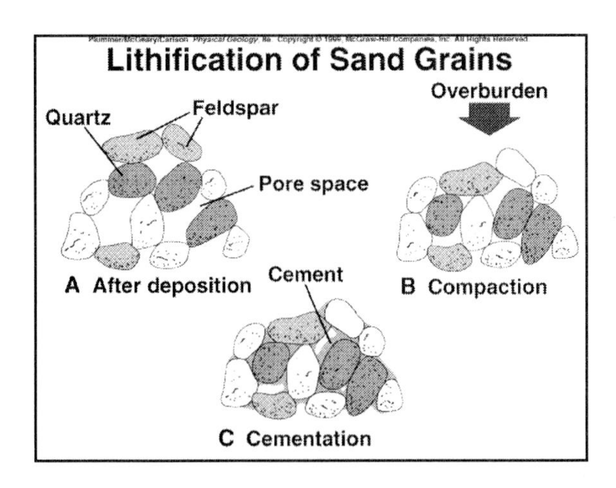
Plummer/McGeary/Carlson Physical Geology, 8e Copyright © 1999, McGraw-Hill Companies, Inc. All Rights Reserved
Lithification of Sand Grains
Quartz
Feldspar
Overburden
Pore space
A After deposition
Cement
B Compaction
C Cementation

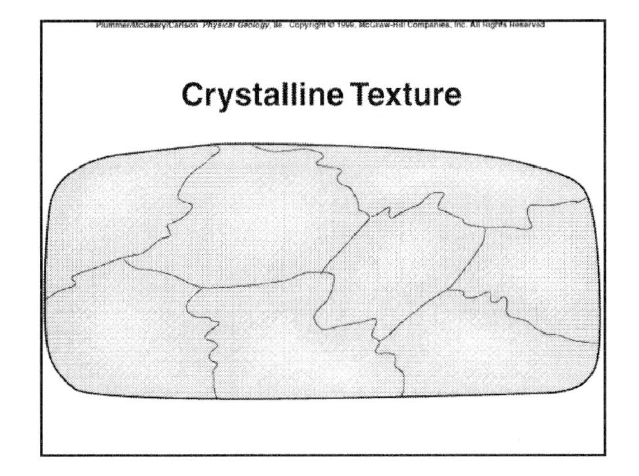
Plummer/McGeary/Carlson Physical Geology, 8e Copyright © 1999, McGraw-Hill Companies, Inc. All Rights Reserved
Crystalline Texture

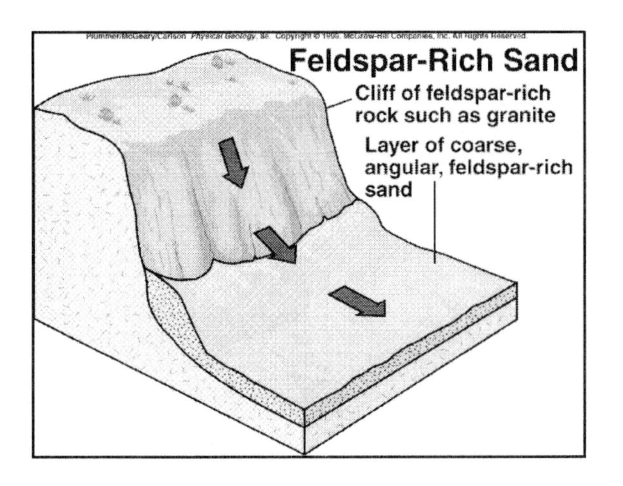
Plummer/McGeary/Carlson Physical Geology, 8e Copyright © 1999, McGraw-Hill Companies, Inc. All Rights Reserved
Feldspar-Rich Sand
Cliff of feldspar-rich rock such as granite
Layer of coarse, angular, feldspar-rich sand

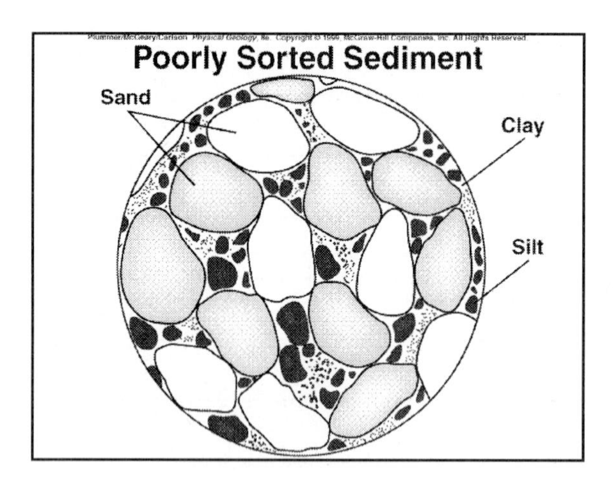
Poorly Sorted Sediment
Sand
Clay
Silt

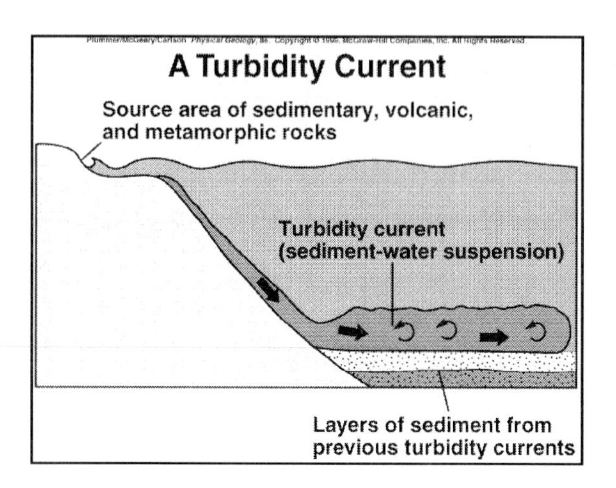
A Turbidity Current
Source area of sedimentary, volcanic, and metamorphic rocks
Turbidity current (sediment-water suspension)
Layers of sediment from previous turbidity currents

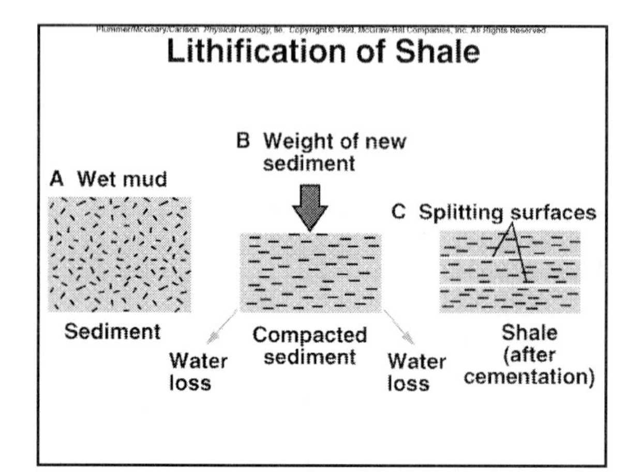
Lithification of Shale
A Wet mud
B Weight of new sediment
C Splitting surfaces
Sediment
Water loss
Compacted sediment
Water loss
Shale (after cementation)

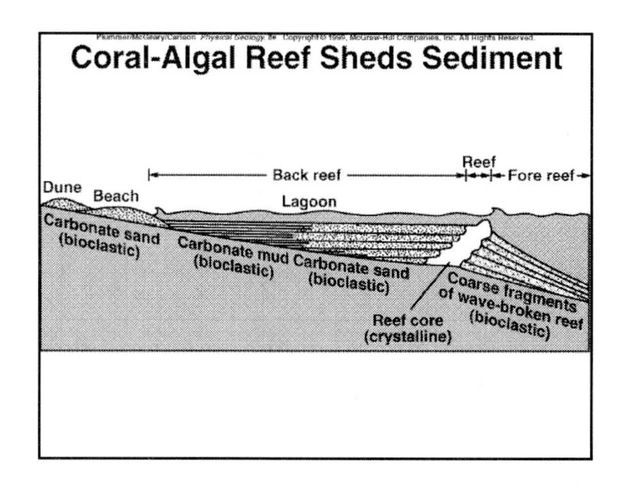
Coral-Algal Reef Sheds Sediment
Back reef
Reef
Fore reef
Dune
Beach
Lagoon
Carbonate sand (bioclastic)
Carbonate mud (bioclastic)
Carbonate sand (bioclastic)
Coarse fragments of wave-broken reef (bioclastic)
Reef core (crystalline)

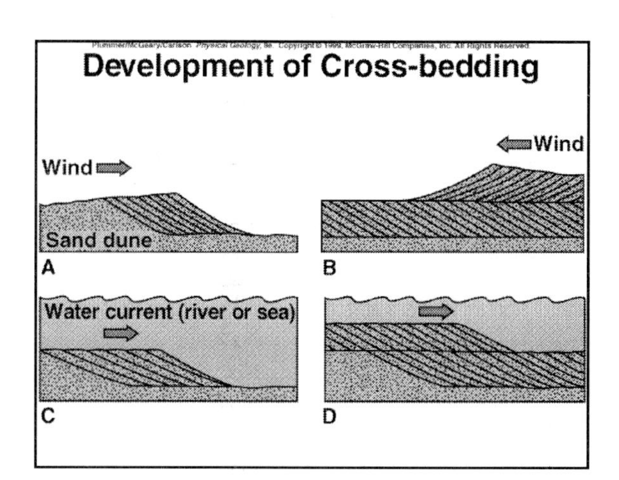
Development of Cross-bedding
Wind
Sand dune
A
Wind
B
Water current (river or sea)
C
D

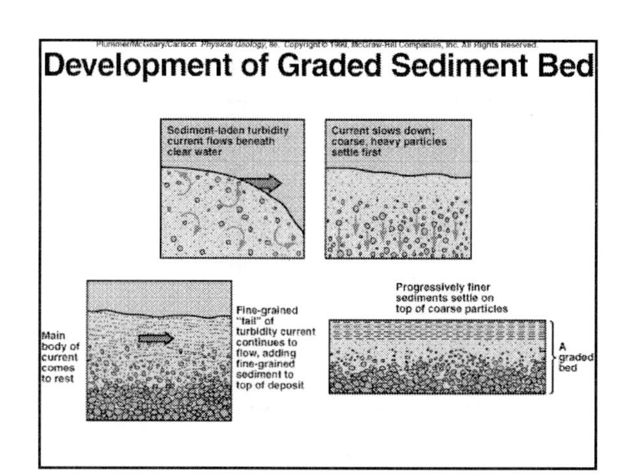
Development of Graded Sediment Bed
Sediment-laden turbidity current flows beneath clear water
Current slows down; coarse, heavy particles settle first
Main body of current comes to rest
Fine-grained "tail" of turbidity current continues to flow, adding fine-grained sediment to top of deposit
Progressively finer sediments settle on top of coarse particles
A graded bed

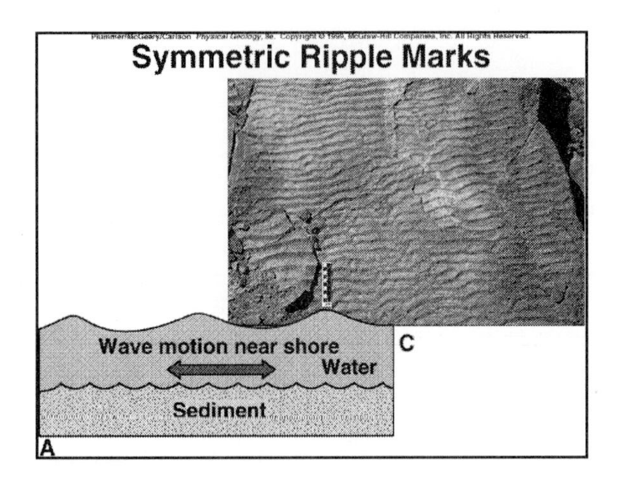
Symmetric Ripple Marks
Wave motion near shore
Water
Sediment
C
A

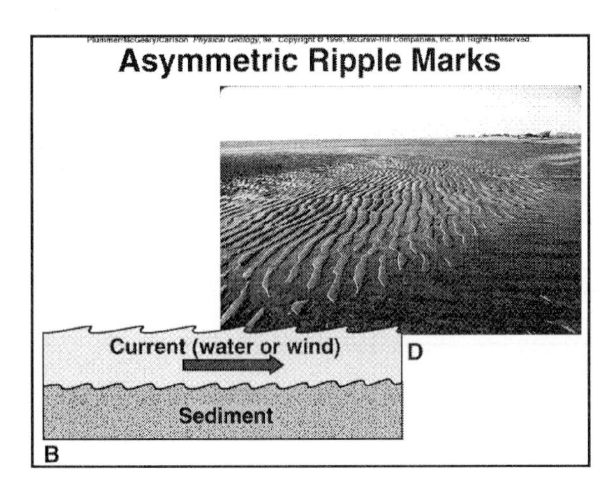
Asymmetric Ripple Marks
Current (water or wind)
Sediment
D
B

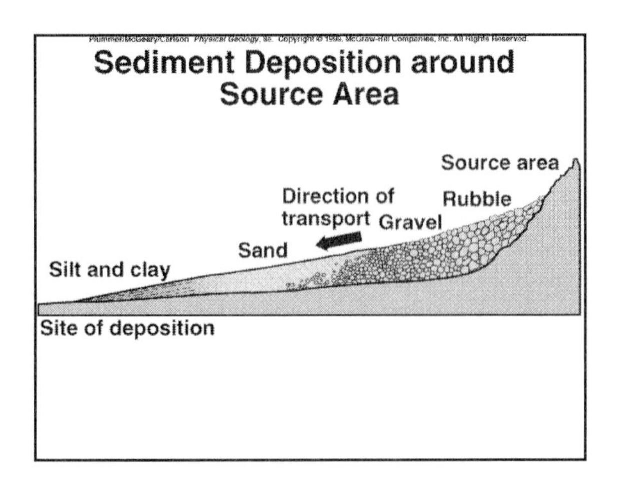
Sediment Deposition around Source Area
Source area
Direction of transport
Rubble
Gravel
Sand
Silt and clay
Site of deposition

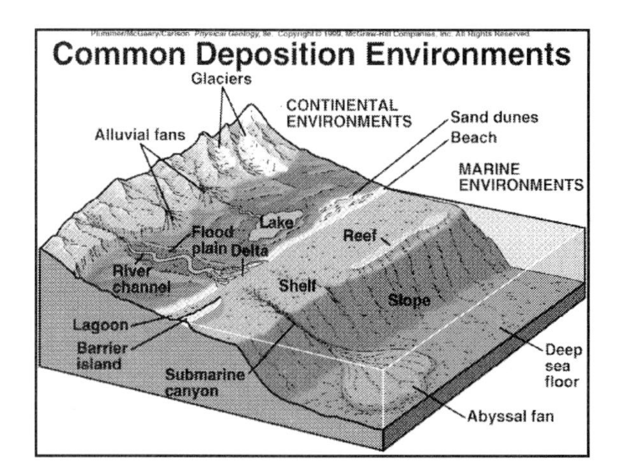
Common Deposition Environments
Glaciers
CONTINENTAL ENVIRONMENTS
Sand dunes
Beach
Alluvial fans
MARINE ENVIRONMENTS
Lake
Flood plain
Delta
Reef
River channel
Shelf
Slope
Lagoon
Barrier island
Submarine canyon
Deep sea floor
Abyssal fan

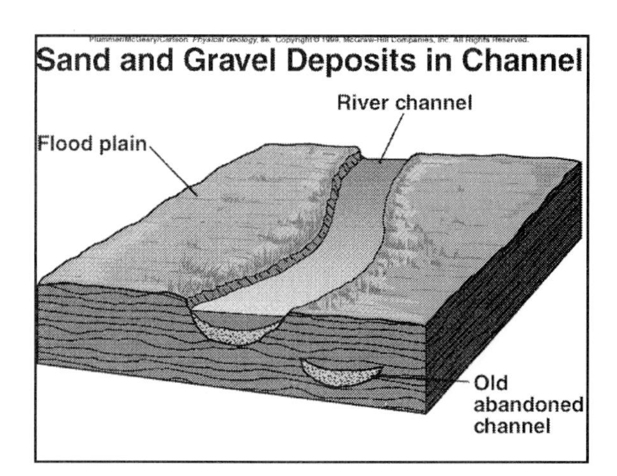
Sand and Gravel Deposits in Channel
River channel
Flood plain
Old abandoned channel

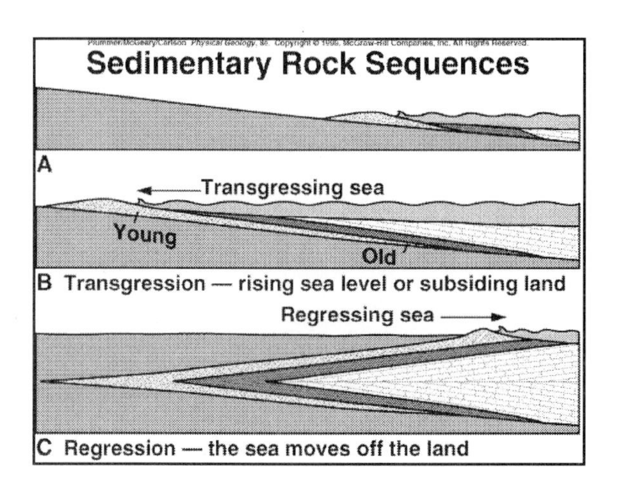
Sedimentary Rock Sequences
A
Transgressing sea
Young
Old
B Transgression — rising sea level or subsiding land
Regressing sea
C Regression — the sea moves off the land

Chapter 7

Physical Geology

eighth edition

Plummer/McGeary/Carlson

METAMORPHISM , METAMORPHIC ROCKS, & HYDROTHERMAL ROCKS

Introduction

- Metamorphism
- Metamorphic rock
- Parent rock
- Equilibrium
- Plastic

Factors Controlling the Characteristics of Metamorphic Rocks

- Composition of Parent Rock
- Temperature
  - Geothermal gradient
  - Stability of minerals
- Pressure
  - Confining pressure
  - Lithostatic pressure

Factors Controlling the Characteristics of Metamorphic Rocks

- Stress
  - Differential Stress
  - Compressive stress
  - Shearing
- Foliation
  - Slaty cleavage
  - Schistose
  - Gneissic texture

Factors Controlling the Characteristics of Metamorphic Rocks

- Fluids
  - Hot water
  - Trigger metamorphic chemical reactions
- Time

Classification of Metamorphic Rocks

- Nonfoliated
  - Named on the basis of its composition
    - Example - quartzite,marble
- Foliated
  - Determine type of foliation
    - Example -schist

Types of Metamorphism

- Contact Metamorphism (thermal metamorphism)
  - Narrow zone of contact
  - Shale => hornfels
  - Limestone => MARBLE
  - Sandstone => QUARTZITE
    - Quartz grains welded together

Types of Metamorphism

- Regional Metamorphism (dynamothermal metamorphism)
  - Takes place at considerable depth and high temperatures
  - Basalt => greenschist- actinolite, and sodium-rich plagioclase
  - Basalt => amphibole schist - hornblende, plagioclase feldspar, and garnet

Types of Metamorphism

- Progressive Metamorphism
  - Progressively greater temperature and pressure
  - Shale => Slate => Phyllite => Schist => Gneiss => Migmatite

Plate Tectonics and Metamorphism

- Gravitational collapse and spreading
- Geothermal gradient
- Isotherm

Hydrothermal Processes

  - Hyrothermal Minerals => Hydrothermal Rock
  - Vein
- Hyrothermal Activity at Diverging Plate Boundaries
- Ore Deposits at Diverging Plate Boundaries
- Metasomatism

Hydrothermal Processes

- Hydothermal Rocks and Minerals
  - disseminated ore deposits
- Sources of Water
  - Groundwater
  - Water Trapped in Sediments
  - Hydrous Minerals

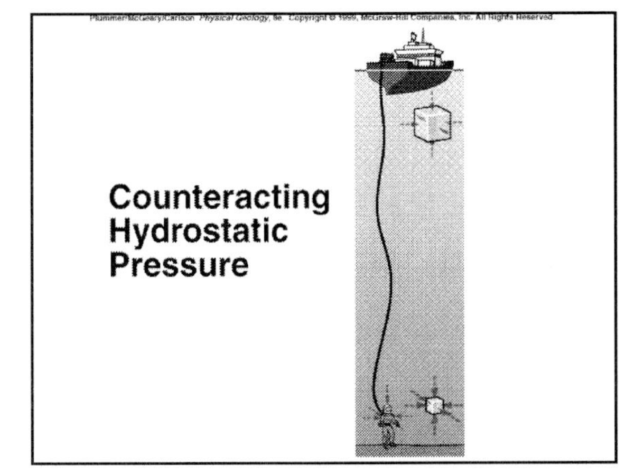

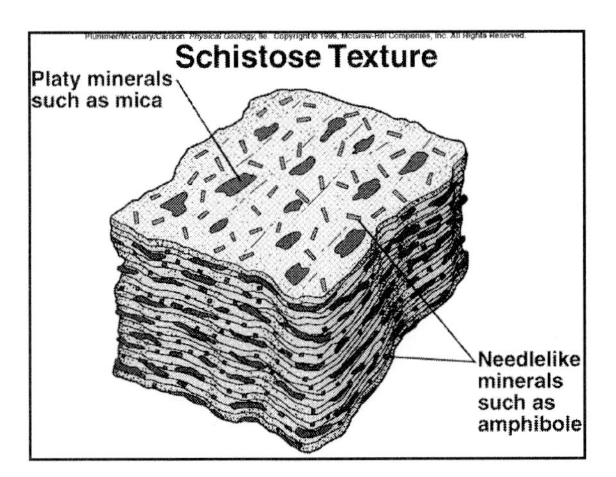

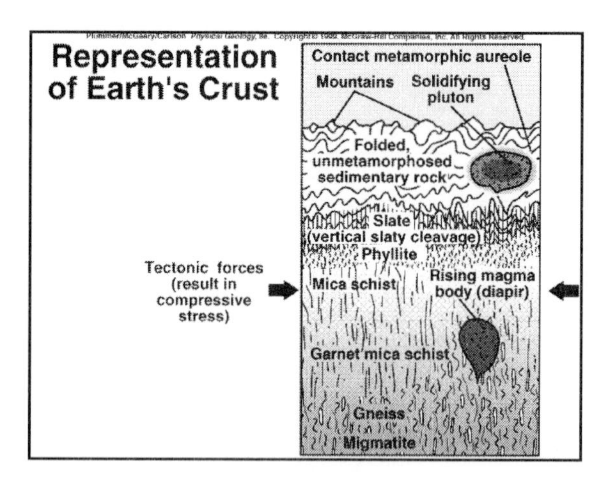
Representation of Earth's Crust
Contact metamorphic aureole
Mountains
Solidifying pluton
Folded, unmetamorphosed sedimentary rock
Slate (vertical slaty cleavage)
Phyllite
Tectonic forces (result in compressive stress)
Mica schist
Rising magma body (diapir)
Garnet mica schist
Gneiss
Migmatite

Slate

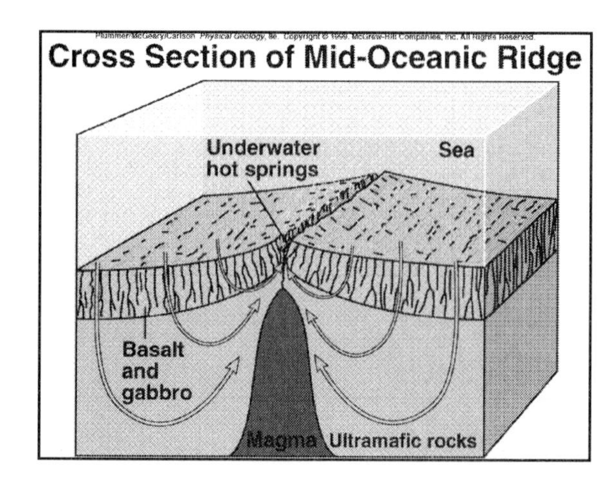
Cross Section of Mid-Oceanic Ridge
Underwater hot springs
Sea
Basalt and gabbro
Magma
Ultramafic rocks

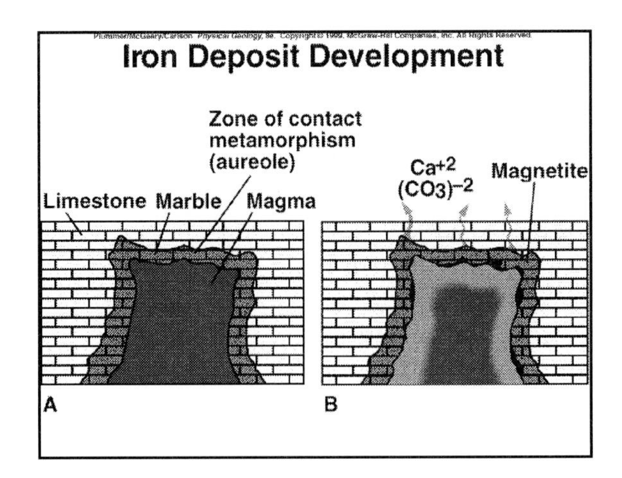
Plummer/McGeary/Carlson Physical Geology, 8e. Copyright © 1999, McGraw-Hill Companies, Inc. All Rights Reserved.
Iron Deposit Development
Zone of contact metamorphism (aureole)
Limestone
Marble
Magma
Ca+2
(CO3)-2
Magnetite
A
B

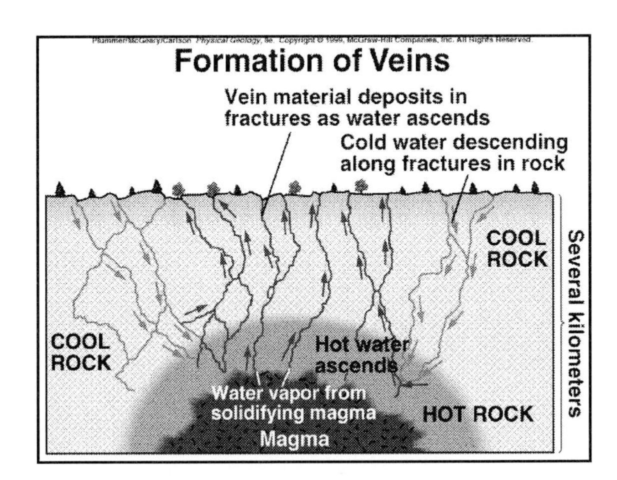
Plummer/McGeary/Carlson Physical Geology, 8e. Copyright © 1999, McGraw-Hill Companies, Inc. All Rights Reserved.
Formation of Veins
Vein material deposits in fractures as water ascends
Cold water descending along fractures in rock
COOL ROCK
COOL ROCK
Hot water ascends
Water vapor from solidifying magma
HOT ROCK
Magma
Several kilometers

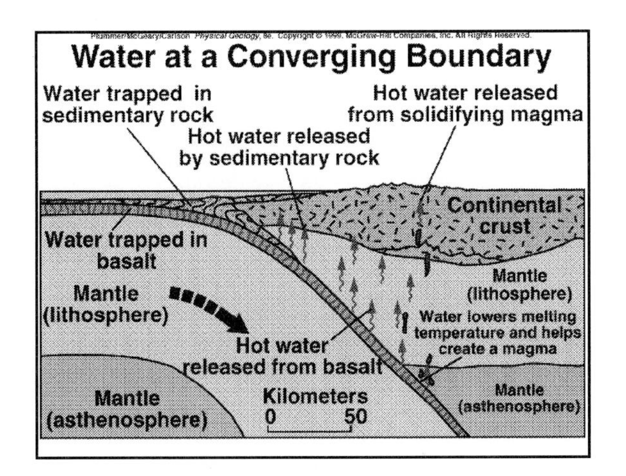
Plummer/McGeary/Carlson Physical Geology, 8e. Copyright © 1999, McGraw-Hill Companies, Inc. All Rights Reserved.
Water at a Converging Boundary
Water trapped in sedimentary rock
Hot water released from solidifying magma
Hot water released by sedimentary rock
Continental crust
Water trapped in basalt
Mantle (lithosphere)
Mantle (lithosphere)
Water lowers melting temperature and helps create a magma
Hot water released from basalt
Mantle (asthenosphere)
Kilometers
0
50
Mantle (asthenosphere)

Chapter 8

Physical Geology

eighth edition

Plummer/McGeary/Carlson

Time and Geology

The Key to the Past

- Geologic Time
- Uniformitarianism
  - uniform rate
- Actualism

- Absolute Age
- Relative Time
  - Sequence of Geologic Events

Principles Used to Determine Relative Age

- Contacts
- Formations

- **Original Horizontality**
- **Superposition**
  - Younger from bottom => top
- **Lateral Continuity**
- **Cross-cutting Relationships**
  - Truncated units

Principles Used to Determine Relative Age

- Other Time Relationships
  - Contact Metamorphism
  - Inclusions

Correlation

- Physical Continuity
- Similarity of Rock Types
- Superposition
- Correlation by Fossils
  - Principle of Faunal Succession
  - Index Fossil
  - Fossil Assemblage

Standard Geologic Time Scale

- Based on Fossil Assemblages
- Eras, Periods, and Epochs
  - Paleozoic Era
  - Mesozoic Era
  - Cenozoic Era
    - Tertiary & Quaternary Periods
      - Recent (Holocene) Epoch
- Precambrian- All time before Paleozoic

Unconformities

- Unconformity - surface that represents a gap in the geologic record
  - Disconformity - contact representing missing parallel beds
  - Angular unconformity - younger strata overlie an erosion surface on tilted or folded layers
  - Nonconformity - erosion surface on plutonic or metamorphic rock

Absolute Age

- Isotopic Dating
- Isotopes and Radioactive Decay
  - Isotopes - Differing number of neutrons, but same number of protons
  - Radioactive Decay - Spontaneous nuclear disintegration of unstable isotopes
    - Daughter Product
    - Half-life
  - Various techniques (e.g. U-Pb, K-Ar, radiocarbon dating)

Absolute Age

- Uses of Isotopic Dating
- Reliability of Isotopic Dating

Combining Relative and Absolute Ages

- Using various principles, as well as-
- Isotopic dating
  - Usually of igneous rocks

Age of the Earth

- Early speculation
- Dating of meteorites
- Between 4.5 & 4.6 billion years old
- Comprehending Geologic Time

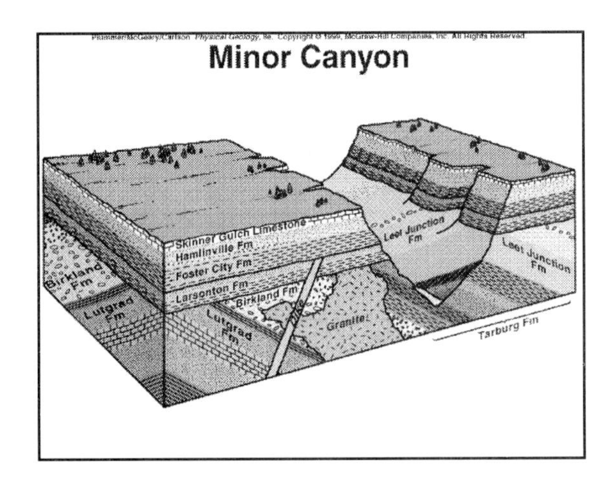
Plummer/McGeary/Carlson Physical Geology, 8e. Copyright © 1999, McGraw-Hill Companies, Inc. All Rights Reserved.
Minor Canyon
Skinner Gulch Limestone
Hamlinville Fm
Foster City Fm
Larsonton Fm
Birkland Fm
Lutgrad Fm
Leet Junction Fm
Granite
Tarburg Fm

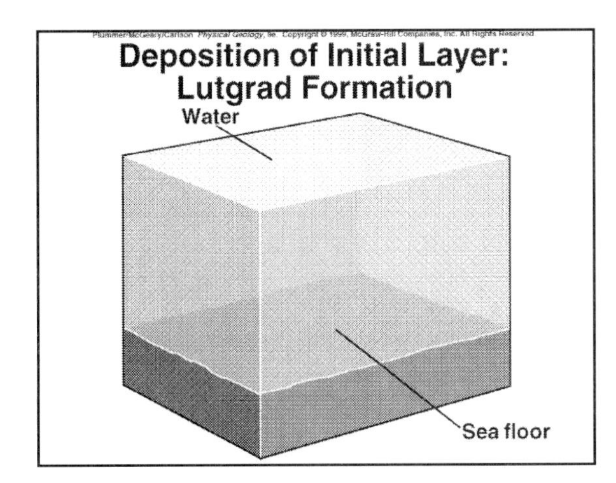
Plummer/McGeary/Carlson Physical Geology, 8e. Copyright © 1999, McGraw-Hill Companies, Inc. All Rights Reserved.
Deposition of Initial Layer:
Lutgrad Formation
Water
Sea floor

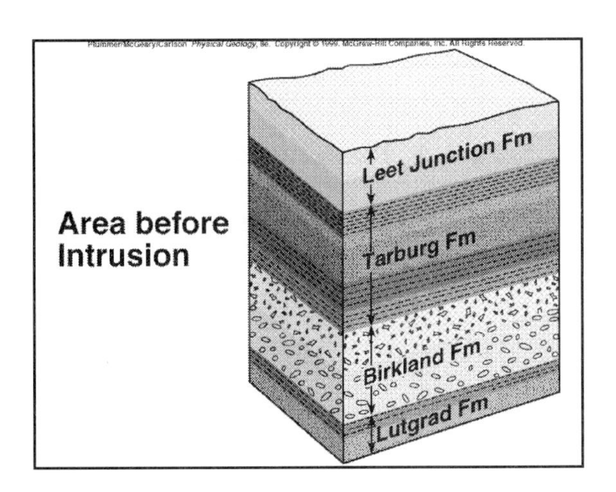
Plummer/McGeary/Carlson Physical Geology, 8e. Copyright © 1999, McGraw-Hill Companies, Inc. All Rights Reserved.
Area before Intrusion
Leet Junction Fm
Tarburg Fm
Birkland Fm
Lutgrad Fm

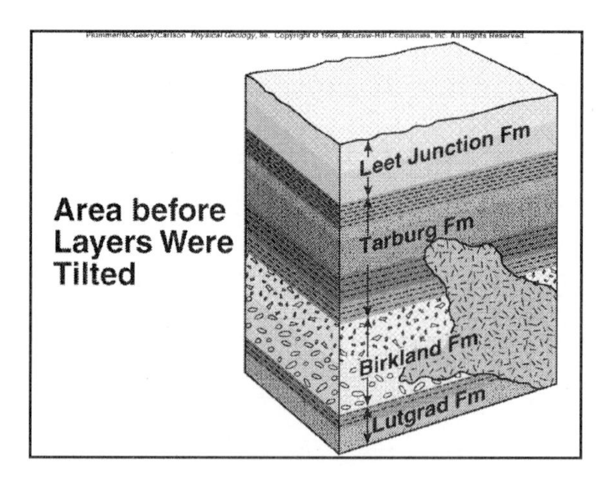
Area before
Layers Were
Tilted
Leet Junction Fm
Tarburg Fm
Birkland Fm
Lutgrad Fm

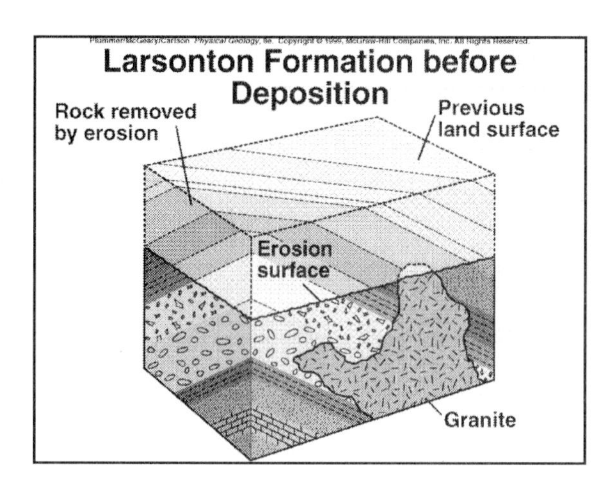
Larsonton Formation before
Deposition
Rock removed
by erosion
Previous
land surface
Erosion
surface
Granite

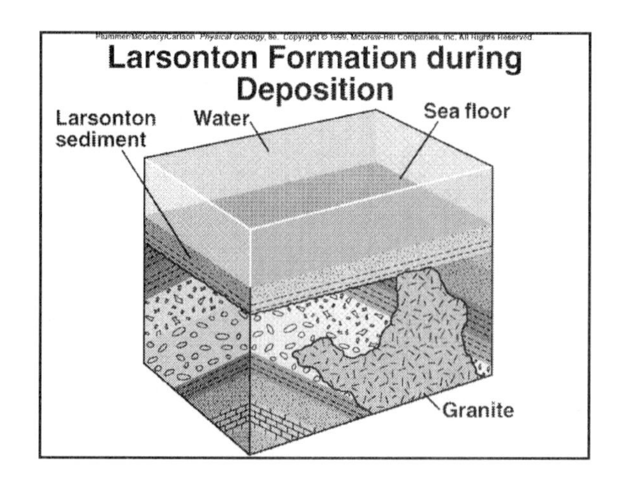
Larsonton Formation during
Deposition
Larsonton
sediment
Water
Sea floor
Granite

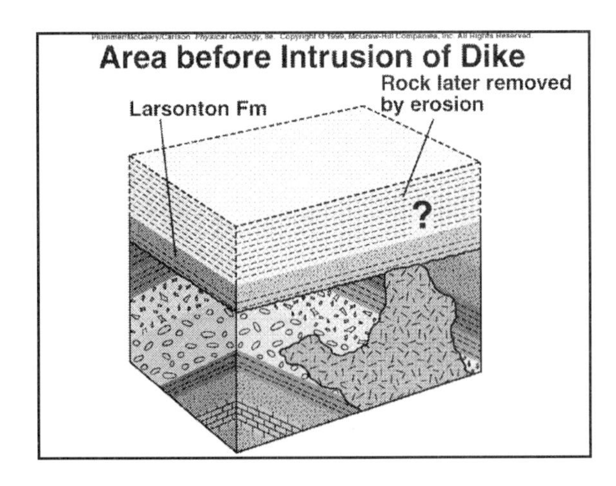
Area before Intrusion of Dike
Rock later removed by erosion
Larsonton Fm
?

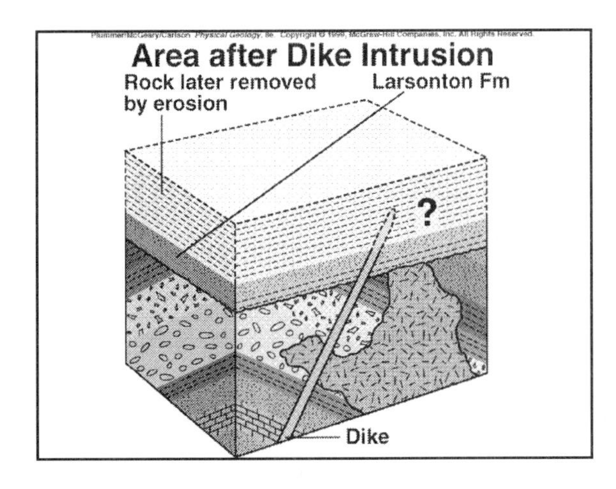
Area after Dike Intrusion
Rock later removed by erosion
Larsonton Fm
?
Dike

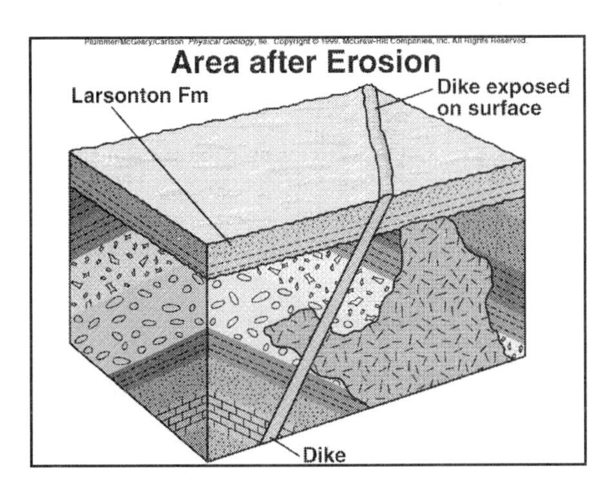
Area after Erosion
Larsonton Fm
Dike exposed on surface
Dike

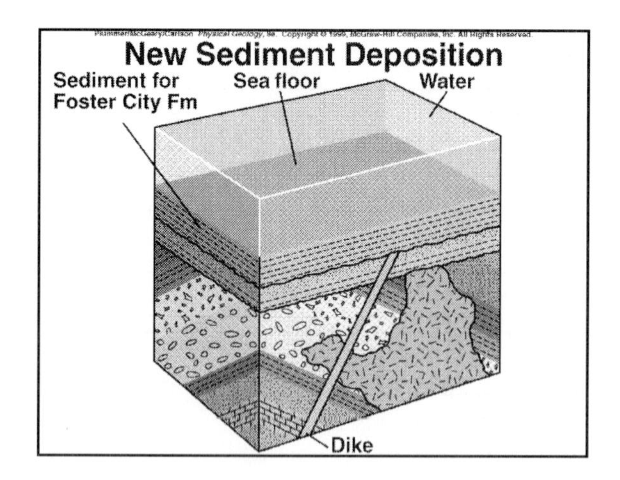
New Sediment Deposition
Sediment for Foster City Fm
Sea floor
Water
Dike

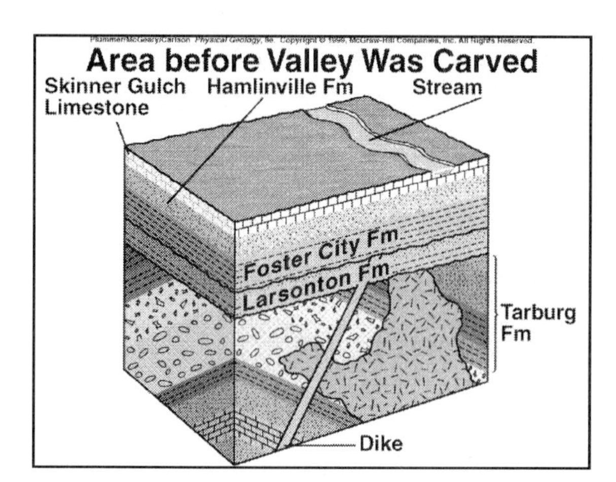
Area before Valley Was Carved
Skinner Gulch Limestone
Hamlinville Fm
Stream
Foster City Fm
Larsonton Fm
Tarburg Fm
Dike

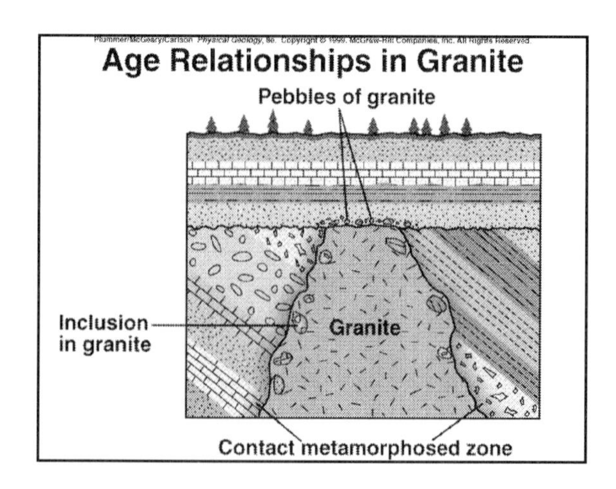
Age Relationships in Granite
Pebbles of granite
Inclusion in granite
Granite
Contact metamorphosed zone

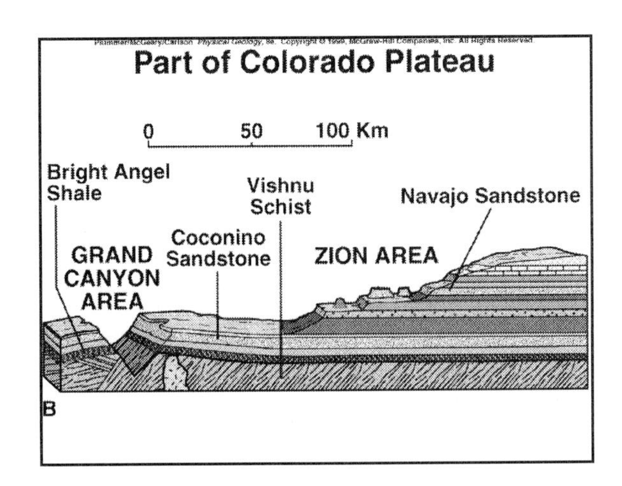

Plummer/McGeary/Carlson Physical Geology, 8e. Copyright © 1999, McGraw-Hill Companies, Inc. All Rights Reserved
Part of Colorado Plateau
0
50
100 Km
Bright Angel Shale
Vishnu Schist
Navajo Sandstone
Coconino Sandstone
GRAND CANYON AREA
ZION AREA
B

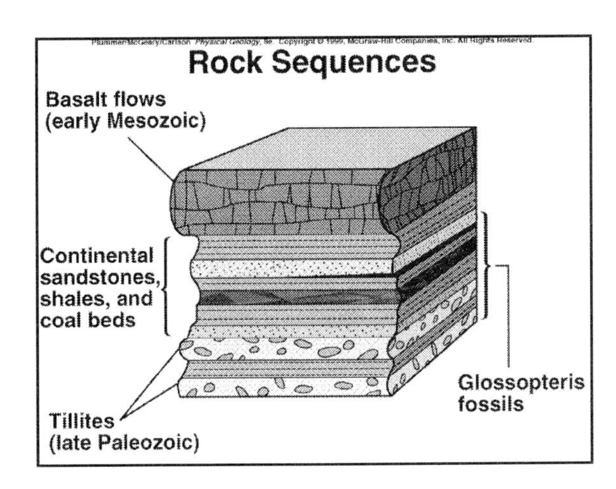

Plummer/McGeary/Carlson Physical Geology, 8e. Copyright © 1999, McGraw-Hill Companies, Inc. All Rights Reserved
Rock Sequences
Basalt flows (early Mesozoic)
Continental sandstones, shales, and coal beds
Tillites (late Paleozoic)
Glossopteris fossils

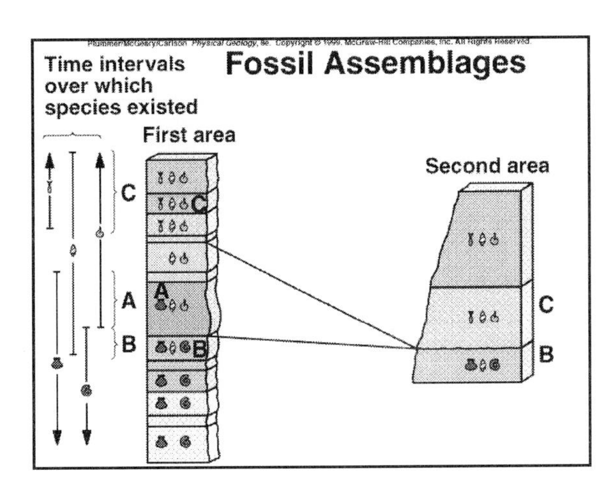

Plummer/McGeary/Carlson Physical Geology, 8e. Copyright © 1999, McGraw-Hill Companies, Inc. All Rights Reserved
Fossil Assemblages
Time intervals over which species existed
First area
Second area
C
A
B
C
A
B
C
B

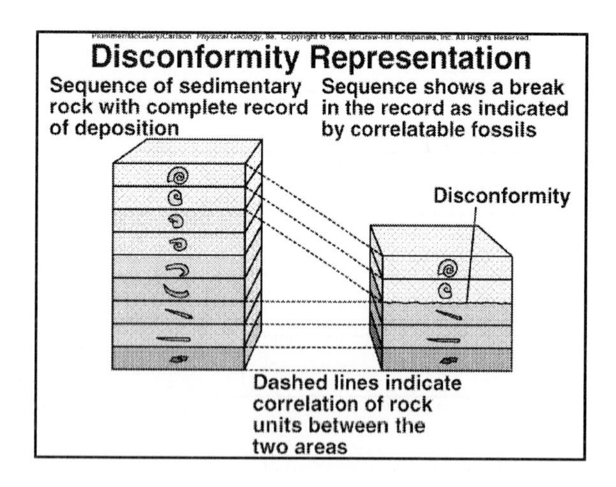
Disconformity Representation
Sequence of sedimentary rock with complete record of deposition
Sequence shows a break in the record as indicated by correlatable fossils
Disconformity
Dashed lines indicate correlation of rock units between the two areas

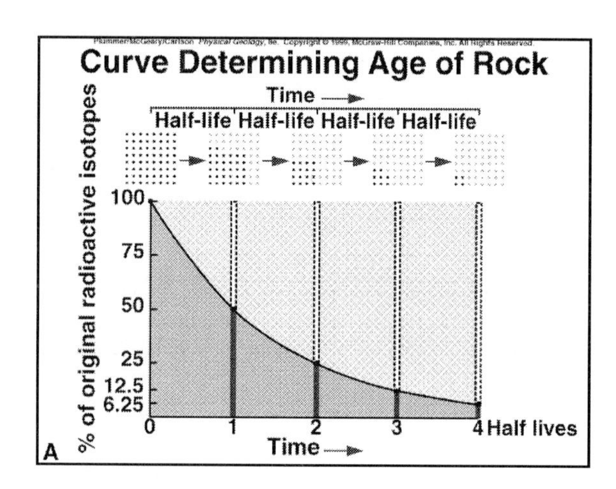
Curve Determining Age of Rock
Time
Half-life
Half-life
Half-life
Half-life
% of original radioactive isotopes
100
75
50
25
12.5
6.25
0
1
2
3
4 Half lives
Time
A

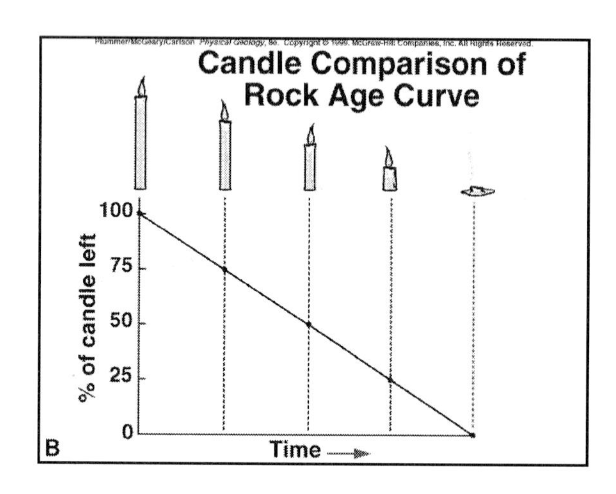
Candle Comparison of Rock Age Curve
% of candle left
100
75
50
25
0
Time
B

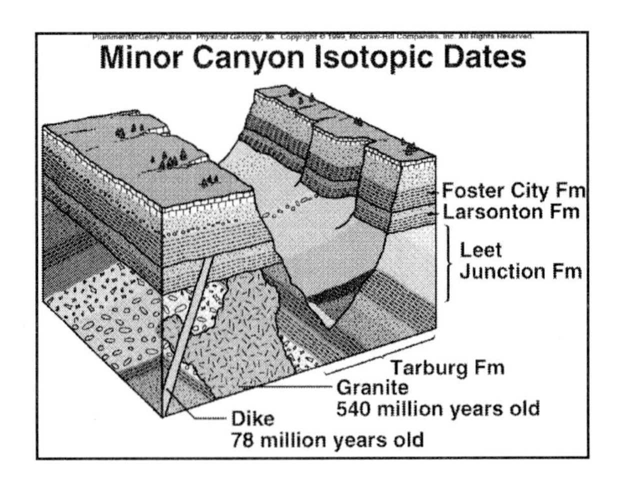
Minor Canyon Isotopic Dates
Foster City Fm
Larsonton Fm
Leet Junction Fm
Tarburg Fm
Granite
540 million years old
Dike
78 million years old

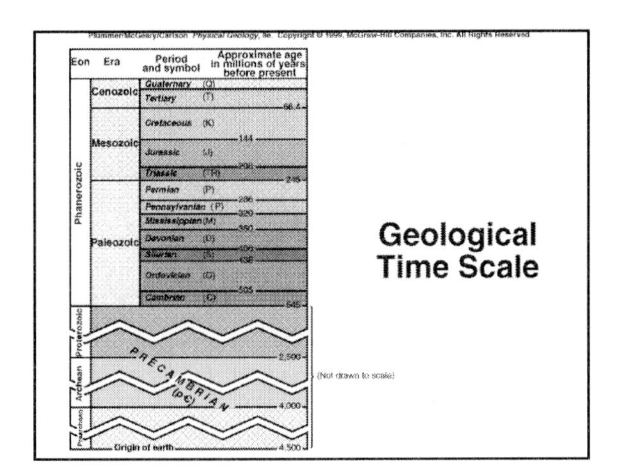
Geological Time Scale
Eon
Era
Period and symbol
Approximate age in millions of years before present
Phanerozoic
Cenozoic
Quaternary (Q)
Tertiary (T)
66.4
Mesozoic
Cretaceous (K)
144
Jurassic (J)
Triassic
245
Paleozoic
Permian (P)
286
Pennsylvanian (P)
320
Mississippian (M)
360
Devonian (D)
Silurian (S)
Ordovician (O)
505
Cambrian (C)
Proterozoic
PRECAMBRIAN (p€)
2,500
Archean
4,000
Origin of earth
4,500
(Not drawn to scale)

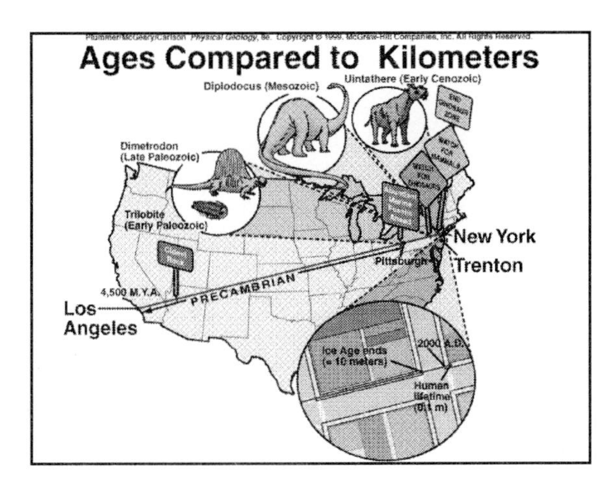
Ages Compared to Kilometers
Diplodocus (Mesozoic)
Uintathere (Early Cenozoic)
Dimetrodon (Late Paleozoic)
Trilobite (Early Paleozoic)
New York
Pittsburgh
Trenton
4,500 M.Y.A.
PRECAMBRIAN
Los Angeles
Ice Age ends (= 10 meters)
2000 A.D.
Human lifetime (0.1 m)

Chapter 9

Physical Geology

eighth edition

Plummer/McGeary/Carlson

MASS WASTING

SURFICIAL PROCESSES

- Erosion, Transportation, Deposition on the Earth's Surface
- Landscapes created and destroyed
- Involves atmosphere, water, gravity
- Agents:
  - Mass wasting, Running water (streams), glaciers, wind, water waves, ground water

MASS WASTING

- Masses of debris or bedrock moving downhill
- Landslides and slower movements
- Driven by GRAVITY

Classification of Mass Wasting

- RATE of MOVEMENT
  - Extremely slow (~1mm/year) to very rapid (>100 km/hour)
- MATERIAL
  - Bedrock
  - Debris- ("soil", sediment)

Classification of Mass Wasting

- TYPE OF MOVEMENT
  - Flow
  - Slide
    - Translational slide
    - Rotational slide (Slump)
  - Fall

Controlling Factors

- Slope angle
- Local relief
- Thickness of debris over bedrock
- Planes of weakness (in bedrock)
  - bedding planes; foliation; joints
  - parallel to slope most dangerous

Controlling Factors

- Climatic controls
  - Ice
  - Water
  - Precipitation
  - Vegetation
- Gravity
  - Shear force
  - Normal force
  - Shear strength

Controlling Factors

- Water
  - adds weight
  - increased pore pressure in saturated debris decreases shear strength
  - surface tension in unsaturated debris increases shear strength
- Triggering Mechanisms
  - Overloading
  - Undercutting
  - Earthquakes

Common types of mass wasting

- CREEP
  - gentle slopes
  - vegetation slows movement
  - very slow *flow* (< 1 cm/year)
    - facilitated by water in soil
    - or by freeze-thaw in colder climates
  - Indicators of creep

Common types of mass wasting

- DEBRIS FLOW
  - Motion taking place throughout moving mass
  - Includes Earthflow, Mudflow, Debris Avalanche

Earthflow

- Primarily *flow* of debris
- may involve *rotational sliding*
- *Scarp* above
- *Hummocky surface* in lower part
- May be slow or fast
- Solifluction
  - role of *Permafrost* in cold climates

Mudflow

- Flow of watery debris
- Occurs where lack of vegetation:
  - Dry climates
  - Volcanoes
  - After forest fires

Debris Avalanche

- Very rapid, turbulent flow of debris

Rockfalls and Rockslides

- **Rockfall**
  - Bedrock breaking loose on cliffs
  - *Talus* at base of cliffs
- **Rockslide**
  - Bedrock involved
  - Sliding along planes of weakness parallel to slope
    - Bedding planes; foliation planes; fractures in rock (joints)

Debris Slides and Debris Falls

- Debris fall
  - Free-falling mass of debris
- Debris slide
  - Debris moving along a well-defined surface

Preventing Landslides

- Preventing mass wasting of debris
- Preventing rockfalls and rockslides on highways

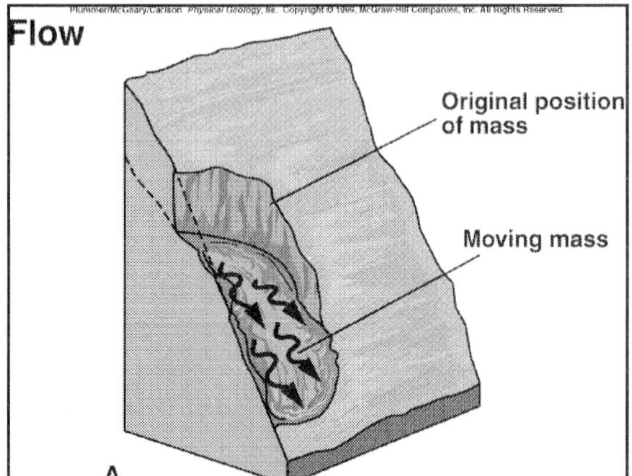

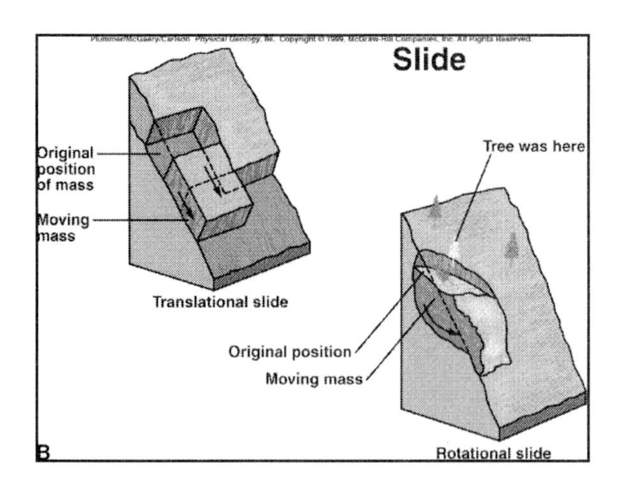

Slide
Original position of mass
Moving mass
Translational slide
Tree was here
Original position
Moving mass
B
Rotational slide

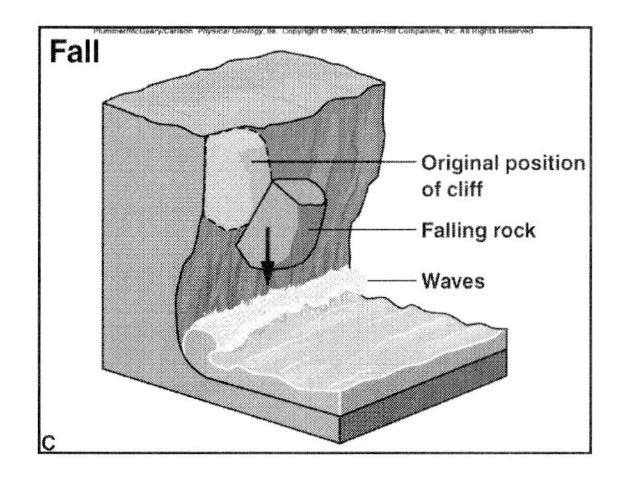

Fall
Original position of cliff
Falling rock
Waves
C

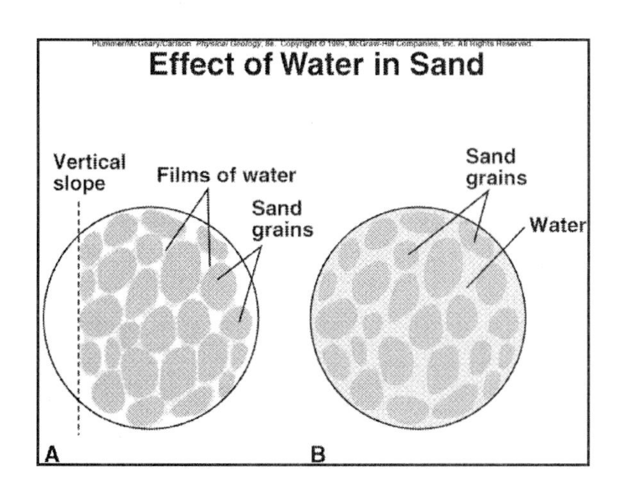

Effect of Water in Sand
Vertical slope
Films of water
Sand grains
A
Sand grains
Water
B

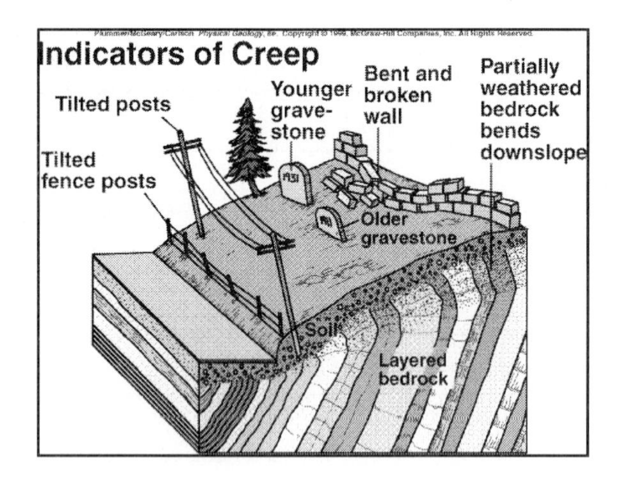
Indicators of Creep
Tilted posts
Tilted fence posts
Younger grave-stone
Bent and broken wall
Partially weathered bedrock bends downslope
Older gravestone
Soil
Layered bedrock

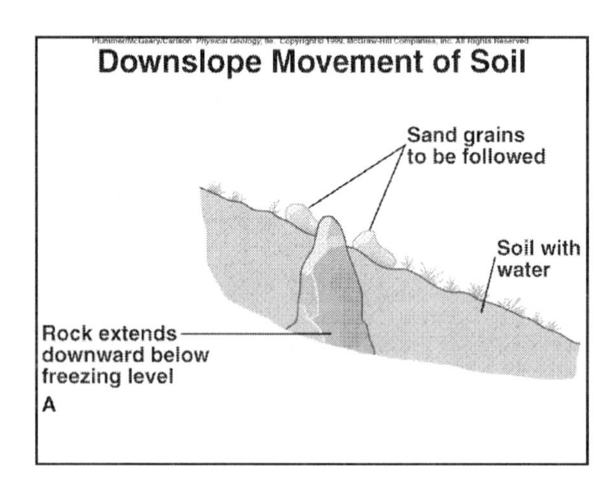
Downslope Movement of Soil
Sand grains to be followed
Soil with water
Rock extends downward below freezing level
A

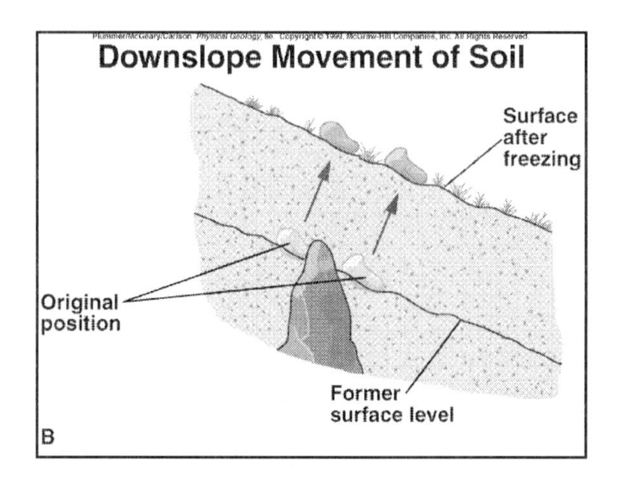
Downslope Movement of Soil
Surface after freezing
Original position
Former surface level
B

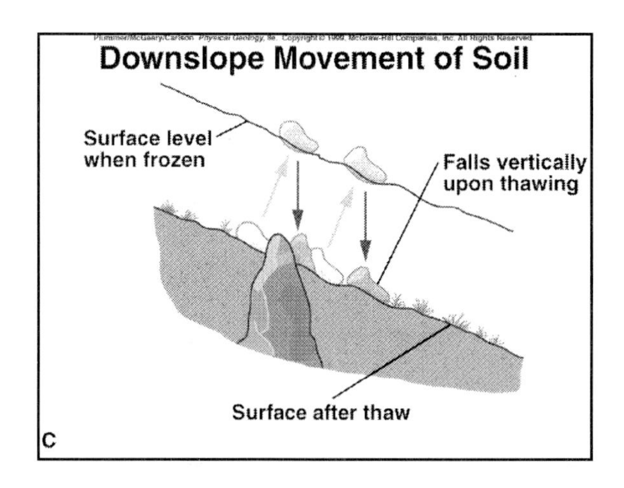
Downslope Movement of Soil
Surface level when frozen
Falls vertically upon thawing
Surface after thaw
C

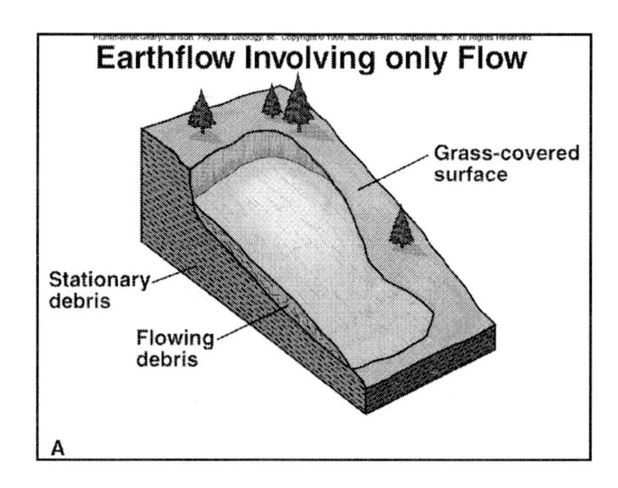
Earthflow Involving only Flow
Grass-covered surface
Stationary debris
Flowing debris
A

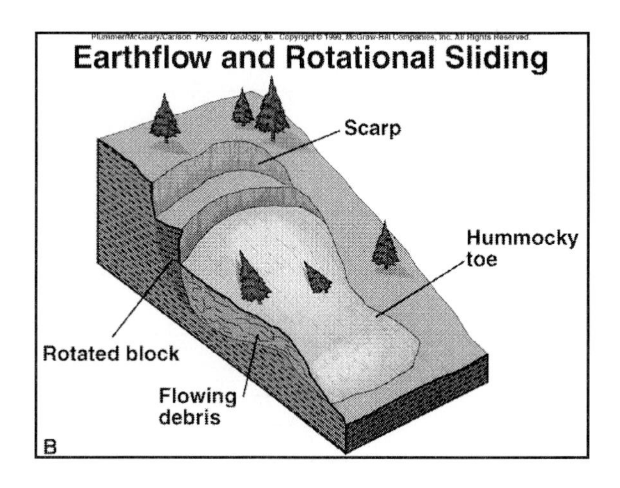
Earthflow and Rotational Sliding
Scarp
Hummocky toe
Rotated block
Flowing debris
B

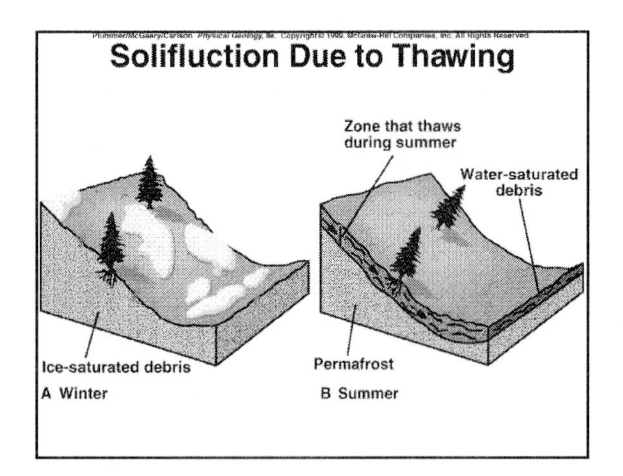
Solifluction Due to Thawing
Zone that thaws during summer
Water-saturated debris
Ice-saturated debris
Permafrost
A Winter
B Summer

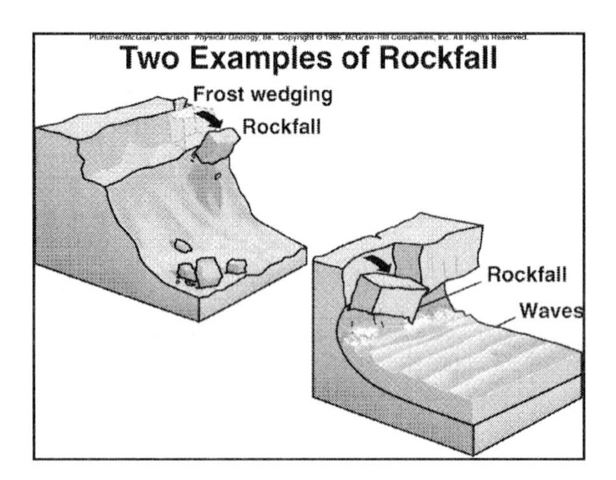
Two Examples of Rockfall
Frost wedging
Rockfall
Rockfall
Waves

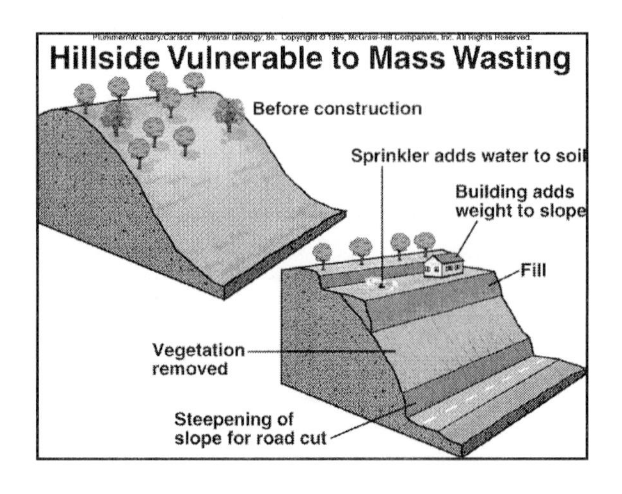
Hillside Vulnerable to Mass Wasting
Before construction
Sprinkler adds water to soil
Building adds weight to slope
Fill
Vegetation removed
Steepening of slope for road cut

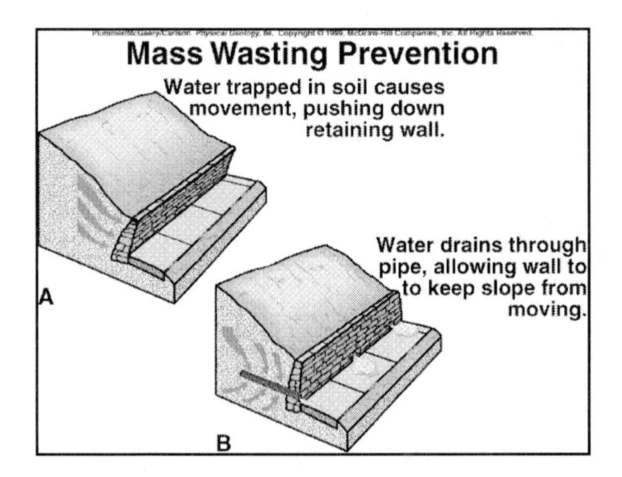
Mass Wasting Prevention
Water trapped in soil causes movement, pushing down retaining wall.
Water drains through pipe, allowing wall to to keep slope from moving.
A
B

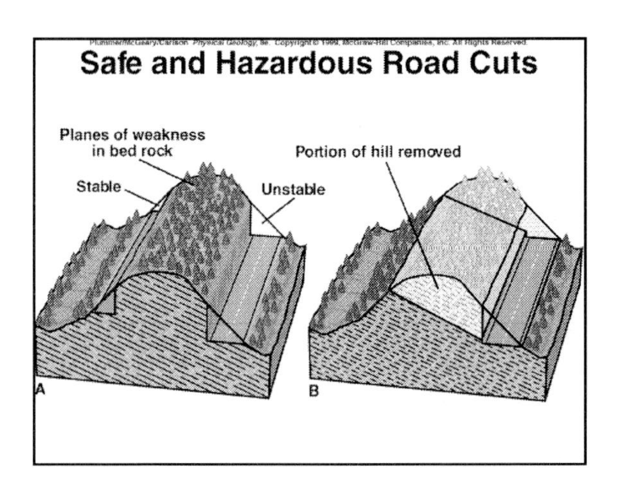
Safe and Hazardous Road Cuts
Planes of weakness in bed rock
Portion of hill removed
Stable
Unstable
A
B

Chapter 10

Physical Geology

eighth edition

Plummer/McGeary/Carlson

Streams and Floods

The Hydrologic Cycle

- What happens to precipitation

Channel Flow and Sheet Flow

- Stream
  - Longitudinal Profile
  - Headwaters; mouth
  - V-shaped cross section
- Floodplain
- Stream channel; bank; bed
- Sheetwash
  - Sheet erosion

Drainage Basins

- Tributary
- Divide

Drainage Patterns

- Dendritic
- Radial
- Rectangular
- Trellis

Factors Affecting Stream Erosion and Deposition

- Velocity
- Gradient
- Channel Shape and Roughness
- Discharge

Stream Erosion

- Hydraulic Action
- Solution
- Abrasion
  - Potholes

Stream Transportation of Sediment

- Bed Load
  - Traction
  - Saltation
- Suspended Load
- Dissolved Load

Stream Deposition

- Bars
  - Placer Deposits
- Braided Streams

Meandering Streams and Point Bars

- Meanders
- Point Bar
- Meander Cutoff
  - Oxbow Lake
- Flood Plains
  - Natural Levees

Deltas & Alluvial Fans

- Distributaries
- Wave Dominated
- Tide Dominated
- Birdfoot Delta
- Stream Dominated
- Bedding Types
  - Foreset Beds
  - Topset Beds
  - Bottomset Beds
- Alluvial Fans

Flooding

- Recurrence Interval
- Flood Erosion
- Flood Deposits

- Urban Flooding
- Flash Floods
- Controlling Floods
- The Great Flood of 1993

Stream Valley Development

- Downcutting and Base Level
- The Concept of a Graded Stream
- Lateral Erosion
- Headward Erosion and Stream Piracy
- Stream Terraces
- Incised Meanders
- Superposed Streams

The Hydrologic Cycle

Solar radiation
Condensation
Transpiration
Precipitation
Evaporation
Evaporation
Runoff in streams
Percolation
Sea
Ground water
Salt water

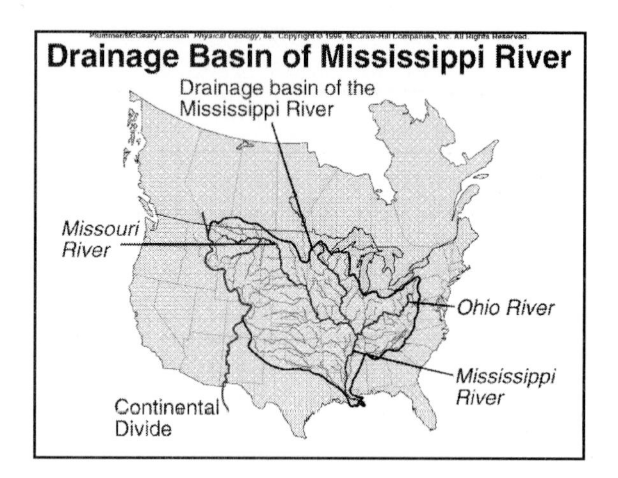
Drainage Basin of Mississippi River
Drainage basin of the Mississippi River
Missouri River
Ohio River
Mississippi River
Continental Divide

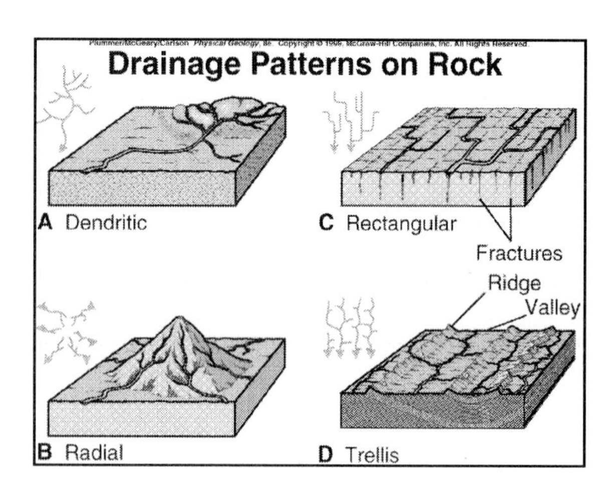
Drainage Patterns on Rock
A Dendritic
C Rectangular
Fractures
Ridge
Valley
B Radial
D Trellis

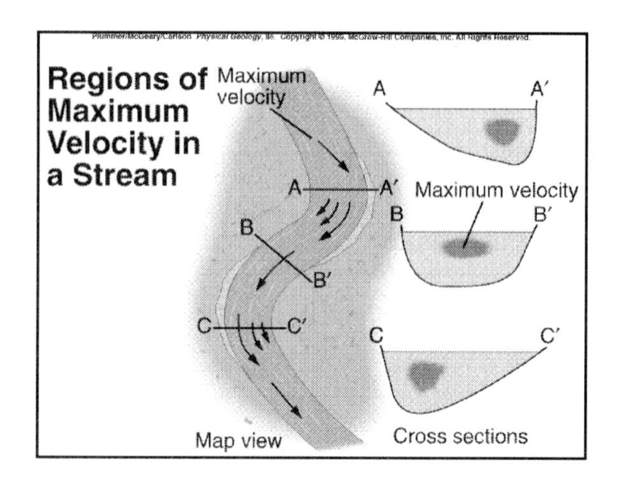
Regions of Maximum Velocity in a Stream
Maximum velocity
A
A′
B
B′
C
C′
Maximum velocity
Map view
Cross sections

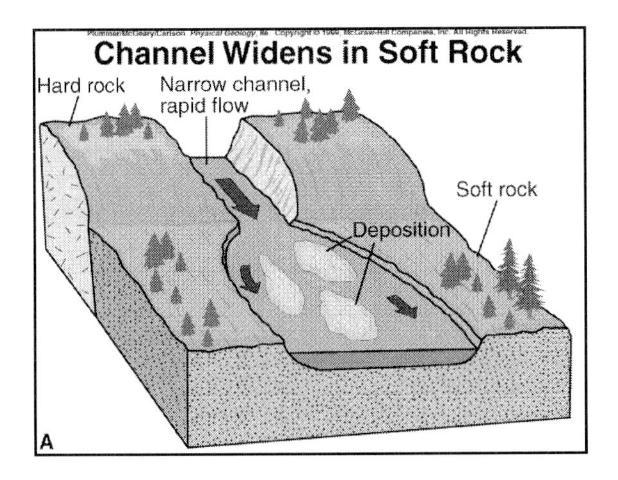
Plummer/McGeary/Carlson Physical Geology, 8e. Copyright © 1999, McGraw-Hill Companies, Inc. All Rights Reserved.
Channel Widens in Soft Rock
Hard rock
Narrow channel,
rapid flow
Soft rock
Deposition
A

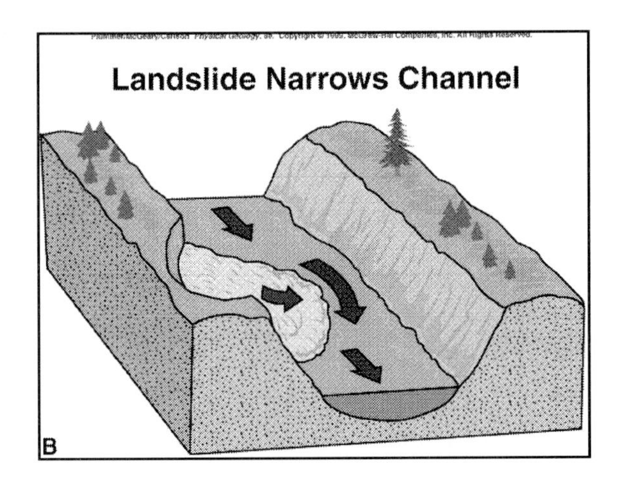
Landslide Narrows Channel
B

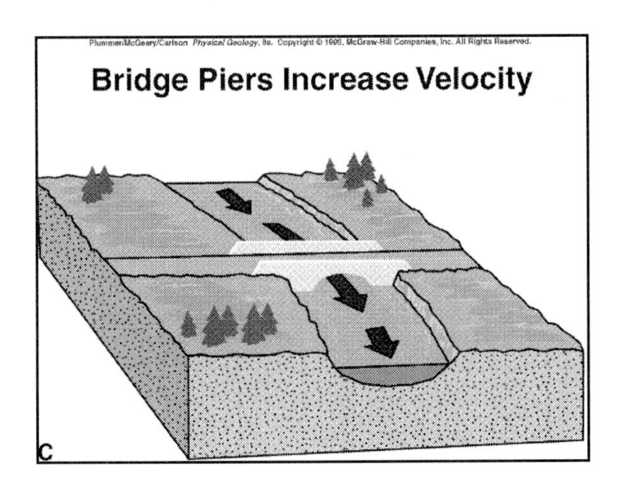
Plummer/McGeary/Carlson Physical Geology, 8e. Copyright © 1999, McGraw-Hill Companies, Inc. All Rights Reserved.
Bridge Piers Increase Velocity
C

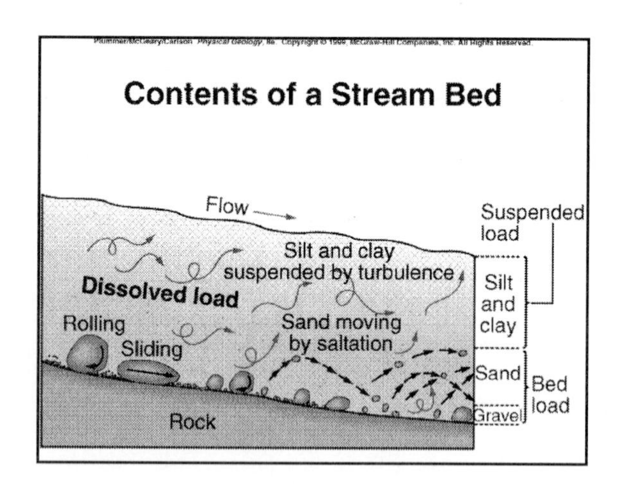
Plummer/McGeary/Carlson Physical Geology, 8e. Copyright © 1999, McGraw-Hill Companies, Inc. All Rights Reserved.
Contents of a Stream Bed
Flow
Suspended load
Silt and clay suspended by turbulence
Dissolved load
Silt and clay
Rolling
Sliding
Sand moving by saltation
Sand
Gravel
Bed load
Rock

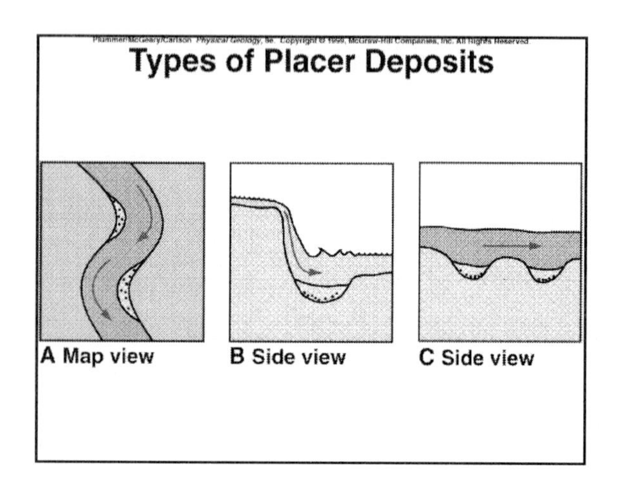
Types of Placer Deposits
A Map view
B Side view
C Side view

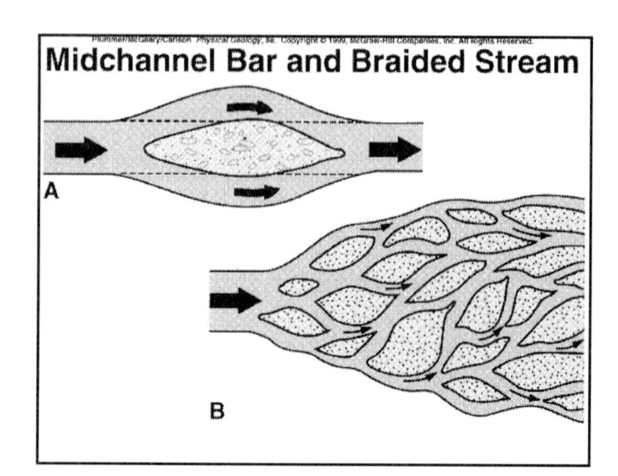
Midchannel Bar and Braided Stream
A
B

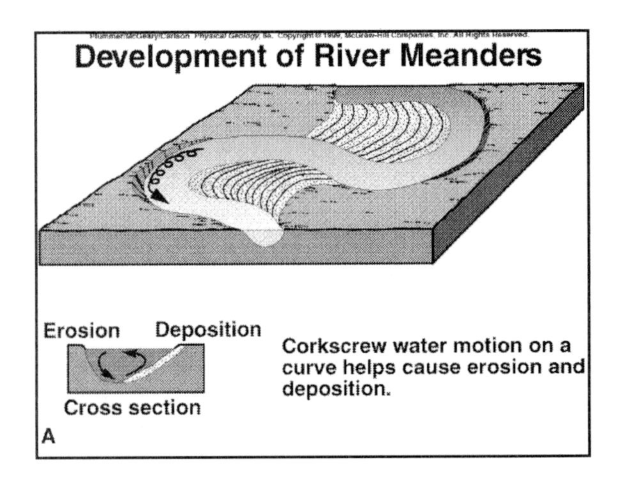
Development of River Meanders
Erosion
Deposition
Cross section
Corkscrew water motion on a curve helps cause erosion and deposition.
A

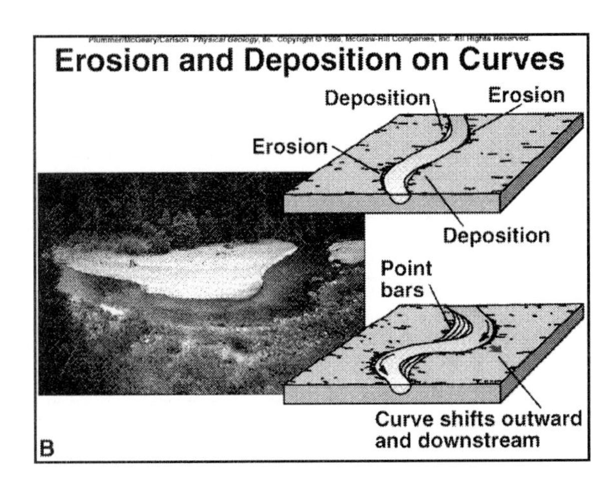
Erosion and Deposition on Curves
Deposition
Erosion
Erosion
Deposition
Point bars
Curve shifts outward and downstream
B

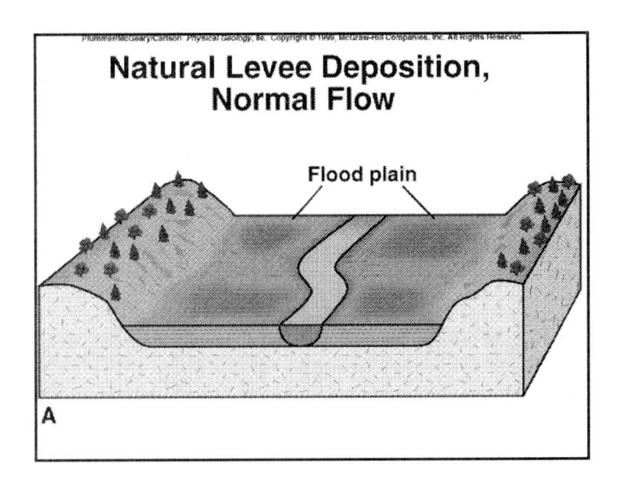
Natural Levee Deposition, Normal Flow
Flood plain
A

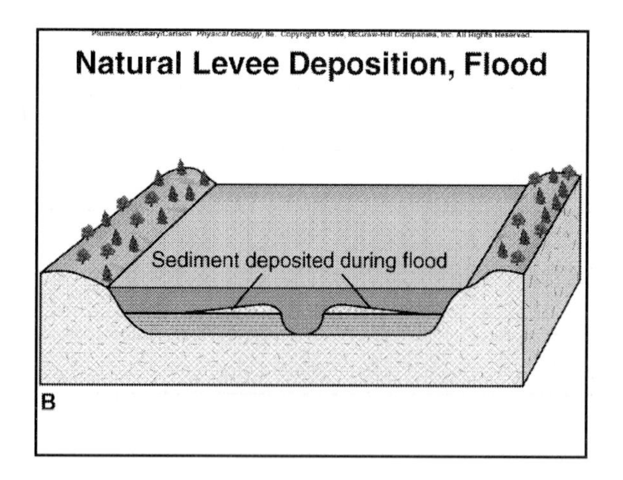
Plummer/McGeary/Carlson Physical Geology, 8e Copyright © 1999, McGraw-Hill Companies, Inc. All Rights Reserved.
Natural Levee Deposition, Flood
Sediment deposited during flood
B

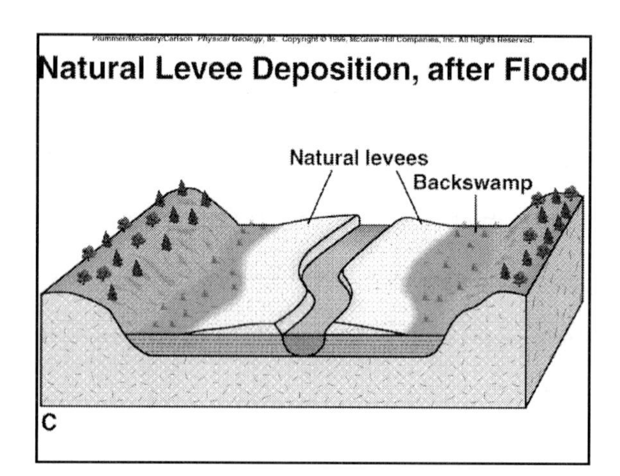
Plummer/McGeary/Carlson Physical Geology, 8e Copyright © 1999, McGraw-Hill Companies, Inc. All Rights Reserved.
Natural Levee Deposition, after Flood
Natural levees
Backswamp
C

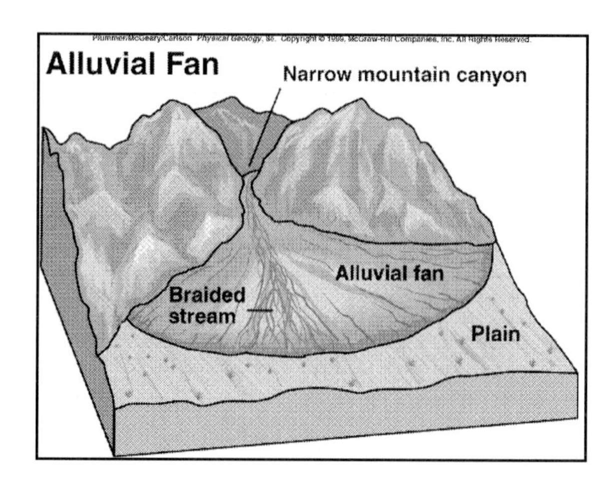
Plummer/McGeary/Carlson Physical Geology, 8e Copyright © 1999, McGraw-Hill Companies, Inc. All Rights Reserved.
Alluvial Fan
Narrow mountain canyon
Alluvial fan
Braided stream
Plain

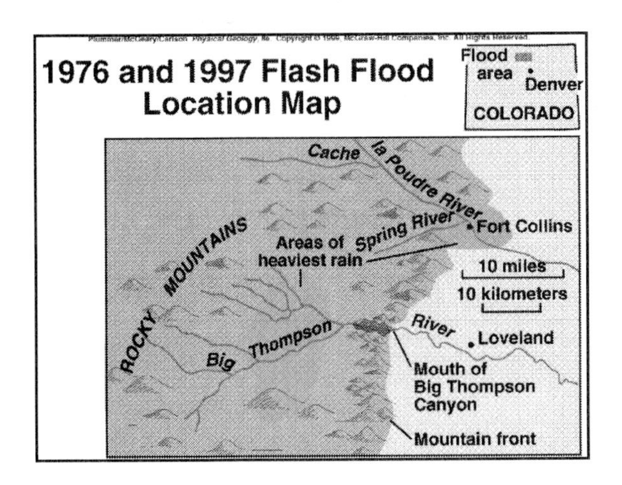
1976 and 1997 Flash Flood Location Map
Flood area
Denver
COLORADO
Cache la Poudre River
ROCKY MOUNTAINS
Areas of heaviest rain
Spring River
Fort Collins
10 miles
10 kilometers
Big Thompson River
Loveland
Mouth of Big Thompson Canyon
Mountain front

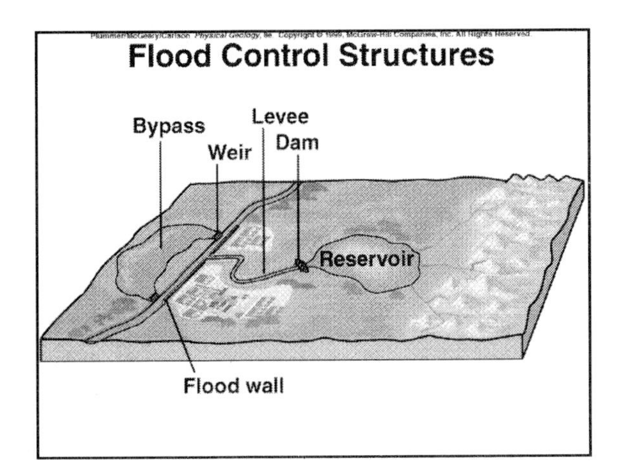
Flood Control Structures
Bypass
Weir
Levee
Dam
Reservoir
Flood wall

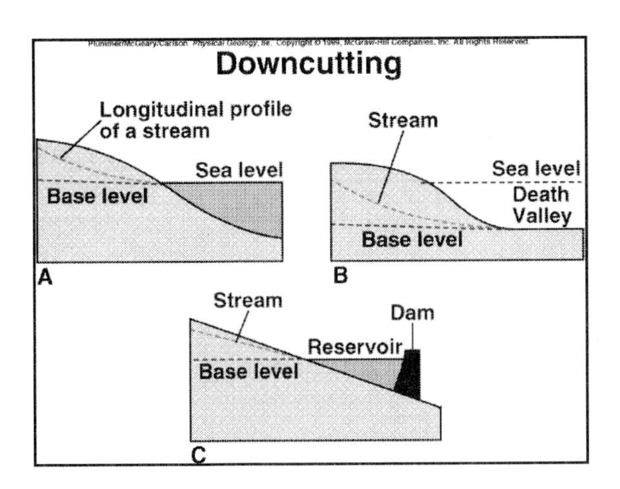
Downcutting
Longitudinal profile of a stream
Sea level
Base level
A
Stream
Sea level
Death Valley
Base level
B
Stream
Dam
Reservoir
Base level
C

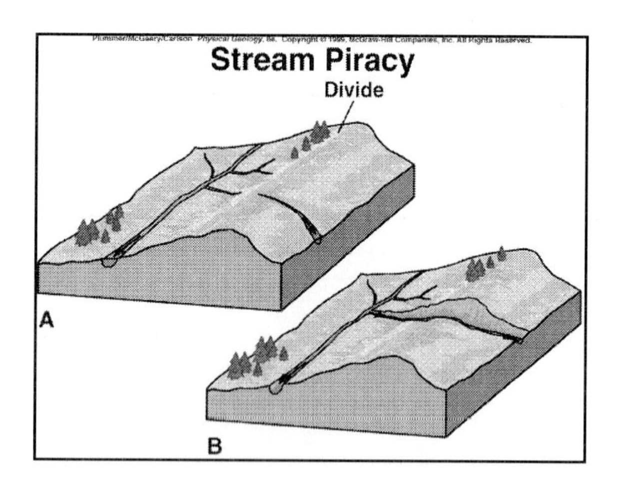
Plummer/McGeary/Carlson Physical Geology, 8e. Copyright © 1999, McGraw-Hill Companies, Inc. All Rights Reserved.
Stream Piracy
Divide
A
B

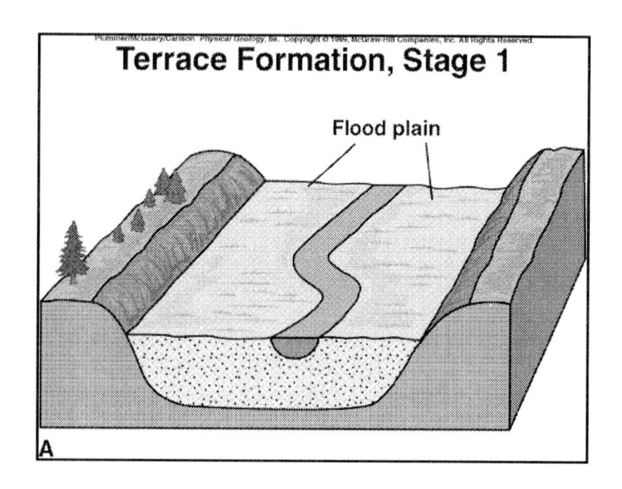
Plummer/McGeary/Carlson Physical Geology, 8e. Copyright © 1999, McGraw-Hill Companies, Inc. All Rights Reserved.
Terrace Formation, Stage 1
Flood plain
A

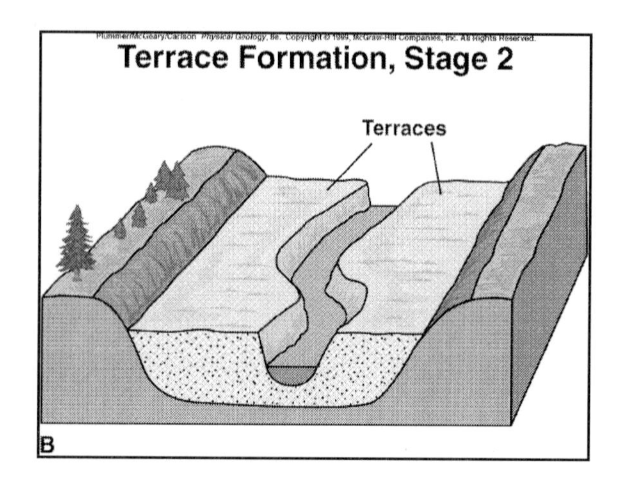
Plummer/McGeary/Carlson Physical Geology, 8e. Copyright © 1999, McGraw-Hill Companies, Inc. All Rights Reserved.
Terrace Formation, Stage 2
Terraces
B

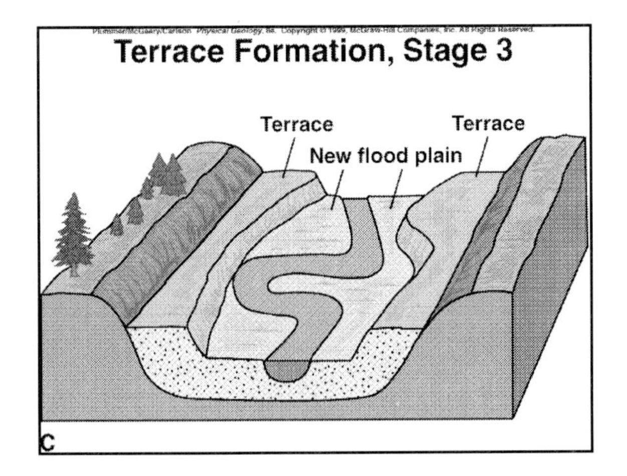
Terrace Formation, Stage 3
Terrace
Terrace
New flood plain
c

Chapter 11

Physical Geology

eighth edition

Plummer/McGeary/Carlson

GROUND WATER

Ground water

- Water filling pore space, cracks & crevices in rocks
- What happens to precipitation

Chapter 12

Physical Geology

eighth edition

Plummer/McGeary/Carlson

GLACIERS AND GLACIATION

GLACIER

- A body of ice
- Formed on land
- Recrystallization of snow
- Evidence of movement
- Alpine glaciation
- Continental glaciation

Theory of glacial ages

- Northern Europe & North America heavily glaciated
  - Peak of glaciation 18,000 years ago

Glaciers-Where they are

- Develop where all of annual snow doesn't melt away in warm seasons
  - Polar regions
  - Heavy winter snowfall
  - High elevations
  - 85% in Antarctica
  - 10% in Greenland

Types of Glaciers

- Valley glacier
- Ice sheet
- Ice cap

Formation and Growth of Glaciers

- Snow to *firn* to *glacier ice*
- Wastage (or ablation)
  - Melting, evaporation,
  - Calving into icebergs
- Glacial Budgets
  - Negative budget- *Receding glacier*
  - Positive budget- *Advancing glacier*
  - Zone of *accumulation;* Zone of *wastage*
    - *Snow line*- divides the zones
  - *Terminus*- movement reflects budget

GLACIERS

- Wastage of glaciers ("shrinkage")
  - Melting
    - more melting at lower elevations
  - Evaporation
  - Calving into Icebergs
    - where a glacier flows onto a sea

GLACIERS

- Advancing vs. Receding Glaciers
- Zone of accumulation
  - Where some snow remains after the melt season
- Zone of Wastage
  - Where all snow & some glacier melt
- Advancing glacier-
  - positive budget - terminus moves forward
- Receding glacier
  - negative budget - terminus retreats

Movement of Glaciers

- Valley Glaciers
  - Gravity driving force
  - Sliding along its base -*basal sliding*
  - Internal flowage- *plastic flow*
  - *Rigid zone*
    - » Crevasses may form here
- Ice sheets
  - Move downward & outward from central high

Glacial Erosion

- Under glacier
  - Abrasion & plucking
  - Bedrock polished & striated
  - Rock flour washes out of glacier
  - Polishing and rounding
    - » "Sheep Rocks"
  - Striations- scratches & grooves on rock
- Above glacier
  - Frost wedging takes place
  - Erosion by glaciers steepens slopes

Erosional Landscapes Associated with Alpine Glaciation

- Glacial valleys
  - U-shaped valleys
  - Hanging valleys
  - Truncated spurs
    - » Triangular facets
  - Rock -basin lakes (tarns)
  - Rounded knobs- *rouche moutonnees*

Erosional Landscapes Associated with Alpine Glaciation

- *Cirque*- at head of valley glacier
  - Rock steps
    - » Rock basin lakes
- Horn
- Arete- sharp ridge

Erosional Landscapes Associated with Continental Glaciation

- Grooved and striated bedrock
  - Grooves may be channels
- Rounded hills & mountains

Glacial Deposition

- Till
  - Unsorted debris
- Erratic
- Moraine- body of till
  - Lateral Moraine
  - Medial Moraine- where tributaries join
  - End moraine-
    - » Terminal
    - » Recessional
  - Ground moraine
  - Drumlin

Glacial Deposition

- Outwash
  - Stream-deposited sediment
    - » sorted
  - Braided streams typical
  - Esker
  - Kettle
- Glacial lakes
  - Varves

Effects of Past Glaciation

- Glacial ages
- Direct effects in North America
  - Scoured much of Canada
  - Cut Great Lakes
  - Deposited till & flattened Midwest
  - Extensive alpine glaciation in mountains

Effects of Glacial Ages

- Indirect effects
  - Pluvial lakes
  - Lowering of sea level
    - » *Fiord*
  - Crustal rebound
- Evidence for older glaciation
  - *Tillite*
  - Late Paleozoic glaciation
    - » Evidence for a supercontinent
  - Precambrian glaciation

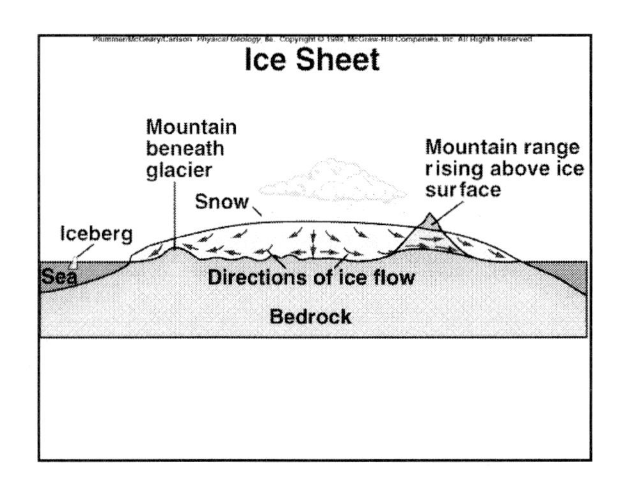
Ice Sheet
Mountain beneath glacier
Mountain range rising above ice surface
Snow
Iceberg
Sea
Directions of ice flow
Bedrock

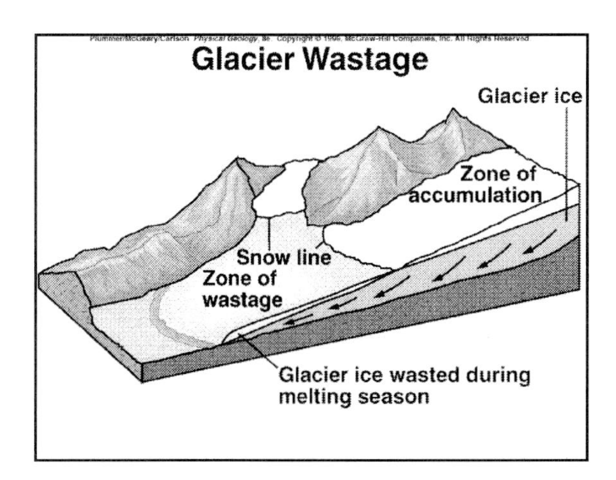
Glacier Wastage
Glacier ice
Zone of accumulation
Snow line
Zone of wastage
Glacier ice wasted during melting season

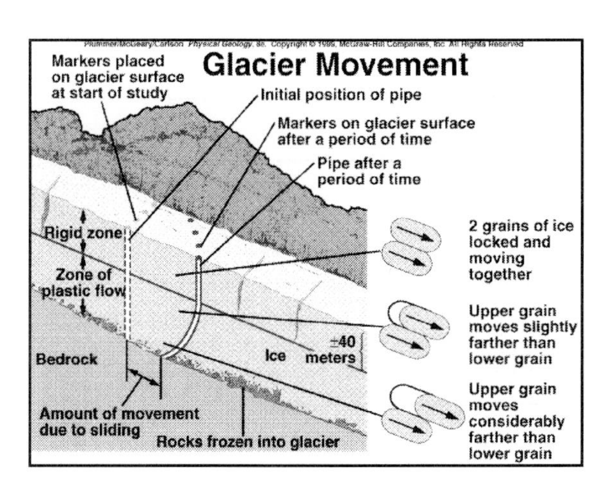
Glacier Movement
Markers placed on glacier surface at start of study
Initial position of pipe
Markers on glacier surface after a period of time
Pipe after a period of time
Rigid zone
Zone of plastic flow
2 grains of ice locked and moving together
Upper grain moves slightly farther than lower grain
±40 meters
Ice
Bedrock
Upper grain moves considerably farther than lower grain
Amount of movement due to sliding
Rocks frozen into glacier

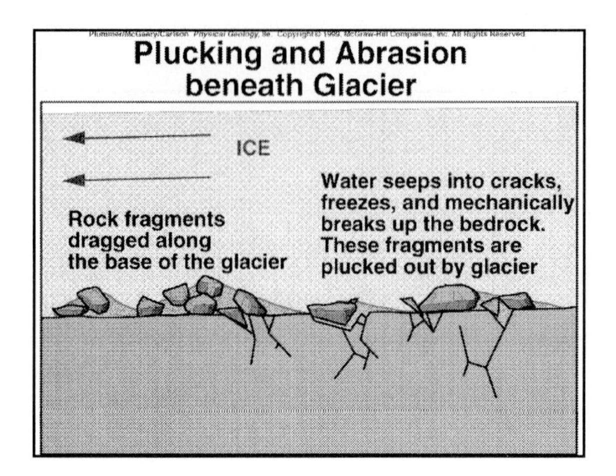
Plucking and Abrasion beneath Glacier
ICE
Rock fragments dragged along the base of the glacier
Water seeps into cracks, freezes, and mechanically breaks up the bedrock. These fragments are plucked out by glacier

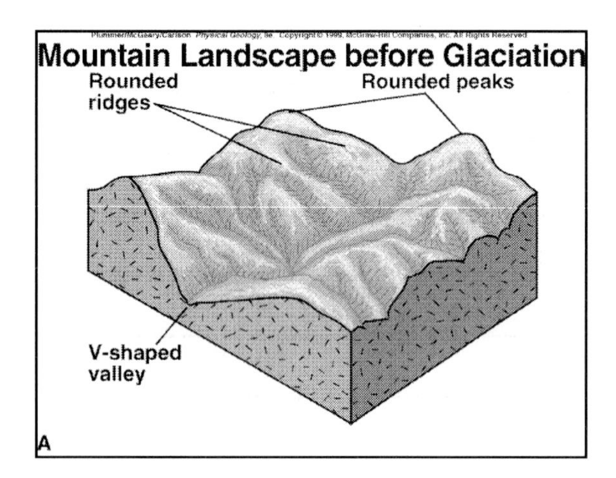
Mountain Landscape before Glaciation
Rounded ridges
Rounded peaks
V-shaped valley
A

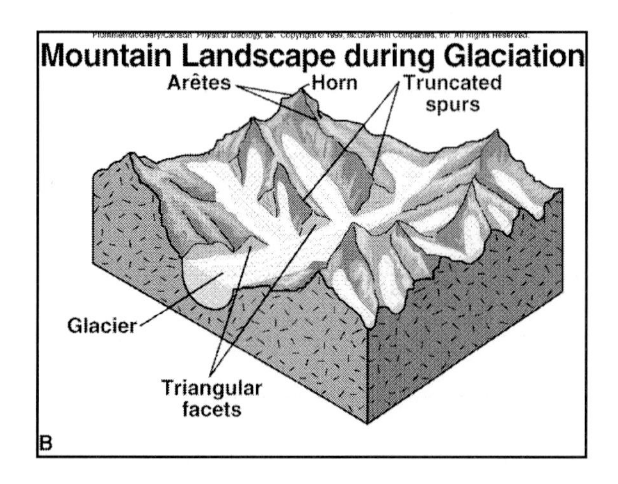
Mountain Landscape during Glaciation
Arêtes
Horn
Truncated spurs
Glacier
Triangular facets
B

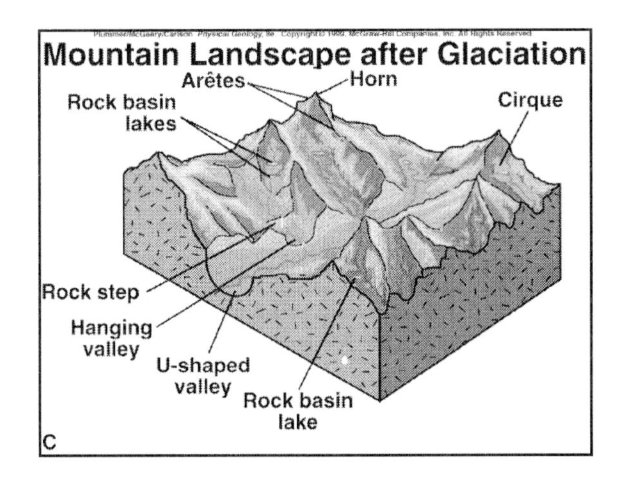
Plummer/McGeary/Carlson Physical Geology, 8e Copyright © 1999, McGraw-Hill Companies, Inc. All Rights Reserved
Mountain Landscape after Glaciation
Arêtes
Horn
Rock basin lakes
Cirque
Rock step
Hanging valley
U-shaped valley
Rock basin lake
C

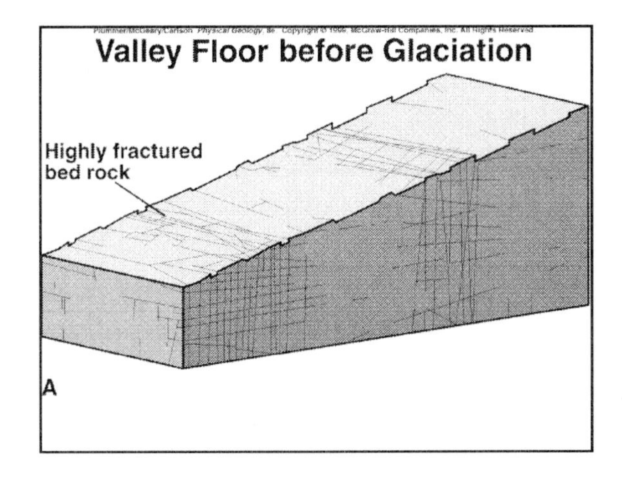
Plummer/McGeary/Carlson Physical Geology, 8e Copyright © 1999, McGraw-Hill Companies, Inc. All Rights Reserved
Valley Floor before Glaciation
Highly fractured bed rock
A

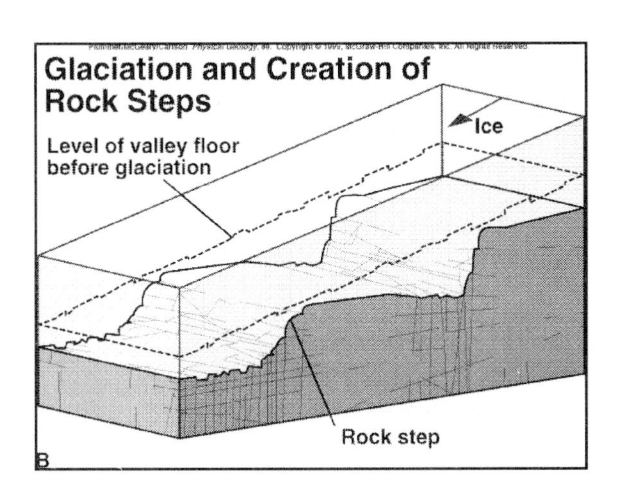
Plummer/McGeary/Carlson Physical Geology, 8e Copyright © 1999, McGraw-Hill Companies, Inc. All Rights Reserved
Glaciation and Creation of Rock Steps
Ice
Level of valley floor before glaciation
Rock step
B

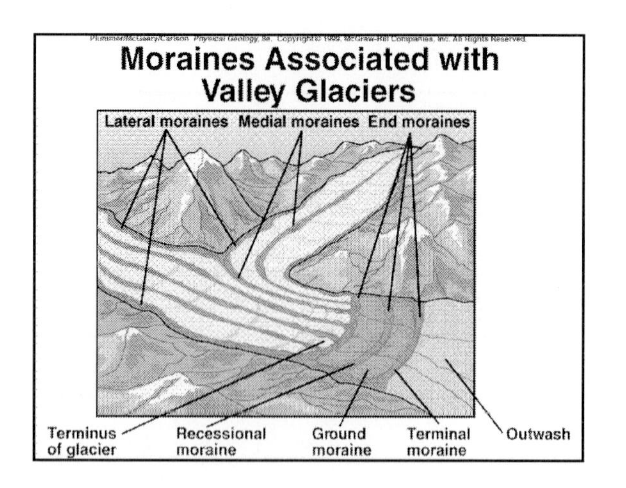
Moraines Associated with Valley Glaciers
Lateral moraines
Medial moraines
End moraines
Terminus of glacier
Recessional moraine
Ground moraine
Terminal moraine
Outwash

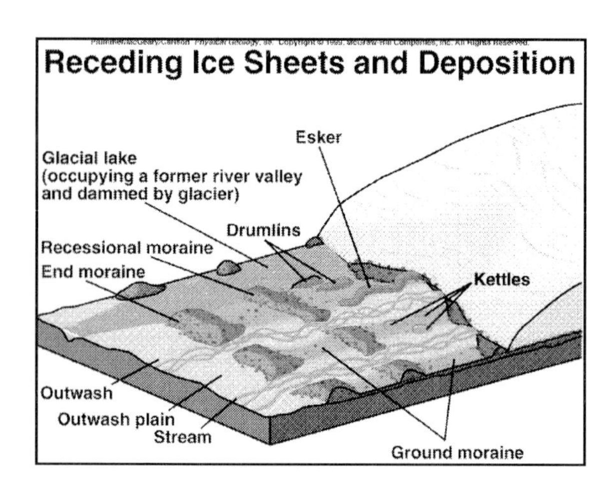
Receding Ice Sheets and Deposition
Esker
Glacial lake (occupying a former river valley and dammed by glacier)
Drumlins
Recessional moraine
End moraine
Kettles
Outwash
Outwash plain
Stream
Ground moraine

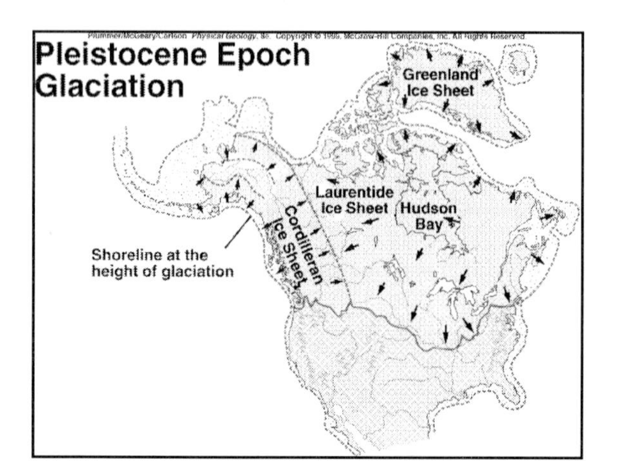
Pleistocene Epoch Glaciation
Greenland Ice Sheet
Laurentide Ice Sheet
Hudson Bay
Cordilleran Ice Sheet
Shoreline at the height of glaciation

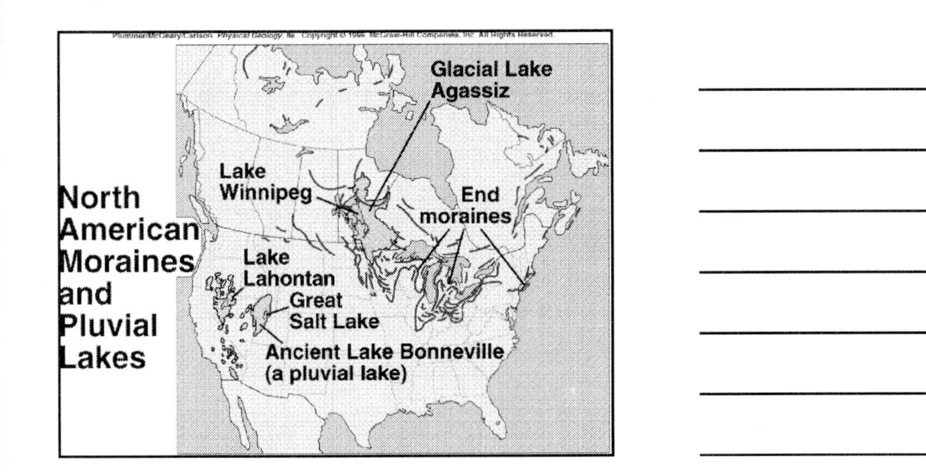
Plummer/McGeary/Carlson Physical Geology 8e Copyright © 1999 McGraw-Hill Companies, Inc. All Rights Reserved.
North American Moraines and Pluvial Lakes
Glacial Lake Agassiz
Lake Winnipeg
End moraines
Lake Lahontan
Great Salt Lake
Ancient Lake Bonneville (a pluvial lake)

Chapter 13

Physical Geology

eighth edition

Plummer/McGeary/Carlson

DESERTS & WIND ACTION

Desert

- Region with low precipitation
  - Usually less than 25 cm of rain per year
- Distribution
  - most related to descending air
  - Belts at 30 degrees North & South latitude
  - Rain shadow of mountains
  - Great distance from oceans
  - Tropical coasts next to cold ocean currents
  - Polar desserts

Some characteristics of deserts

- Lack of through-flowing streams
- Internal drainage
- Local base levels
- Desert thunderstorms
  - Flash floods
    - Mudflows

Some characteristics of deserts

- Stream channels normally dry
  - covered with sand & gravel
  - Narrow canyons with vertical walls
- Resistance of rocks to weathering
  - Desert topography typically steep and angular

Desert Features in S.W. United States

- Colorado Plateau
  - Mostly flat-lying sedimentary beds
  - Plateaus, mesas, buttes
  - Monoclines
    - Hogback; cuesta

Desert Features in S.W. United States

- Basin and Range Province
  - Mountains & valleys bounded by faults
  - Alluvial fans; bajada
  - Playa lake; playa
  - Pediment
  - Parallel retreat of slope

Wind Action

- Strong in desert because:
  - Low humidity
  - Great temperature ranges
  - More effective because of lack of vegetation
- Effective erosion in deserts because sediment is dry

Wind Erosion and Transportation

- Dust storms
- Sand
  - Moves along ground- saltation
  - Sandstorms
  - Sandblasting up to 1 meter
    - Ventifact
- Deflation
  - Blowout

Wind Deposition

- Loess
- Sand Dunes
  - Well-sorted, well-rounded sand grains
  - *Slip face*
    - Angle of repose
  - Wind ripples

Wind Deposition

- Types of dunes
  - Barchan
  - Transverse dune
  - Parabolic dune
  - Longitudinal dune

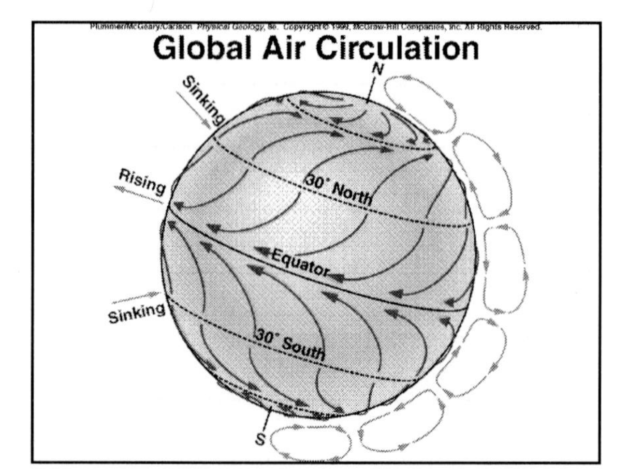

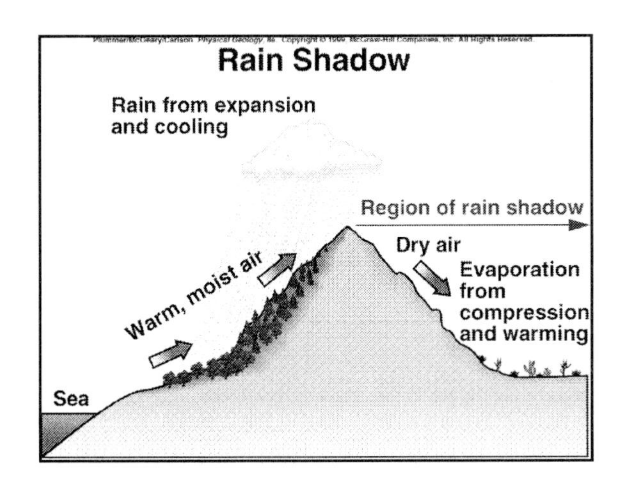
Rain Shadow
Rain from expansion and cooling
Region of rain shadow
Dry air
Warm, moist air
Evaporation from compression and warming
Sea

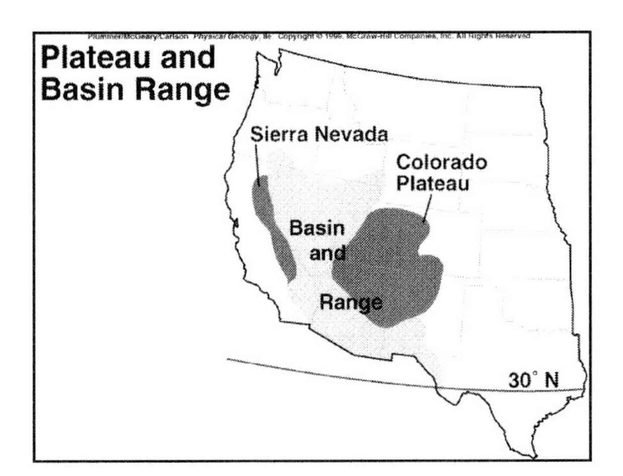
Plateau and Basin Range
Sierra Nevada
Colorado Plateau
Basin and Range
30° N

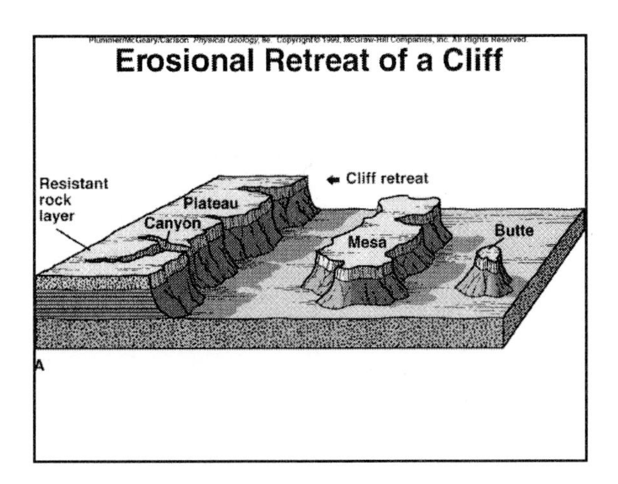
Erosional Retreat of a Cliff
Resistant rock layer
Plateau
Canyon
Cliff retreat
Mesa
Butte
A

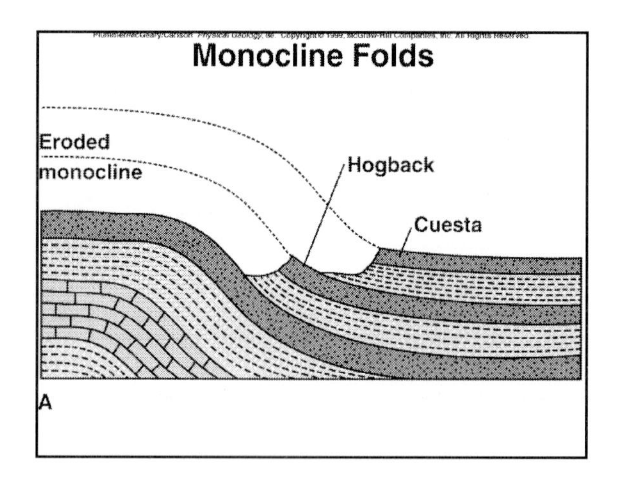
Monocline Folds
Eroded
monocline
Hogback
Cuesta
A

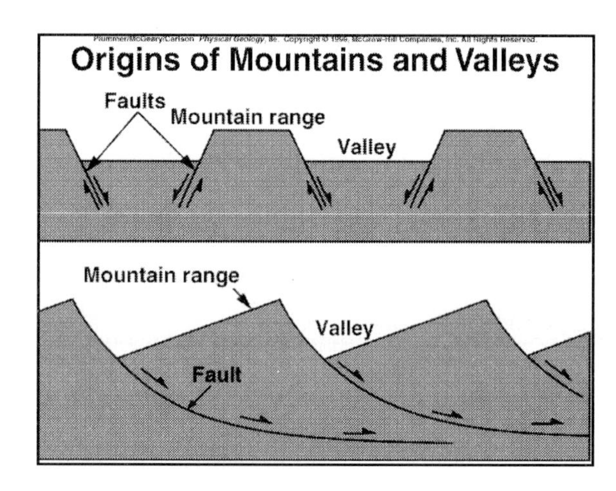
Origins of Mountains and Valleys
Faults
Mountain range
Valley
Mountain range
Valley
Fault

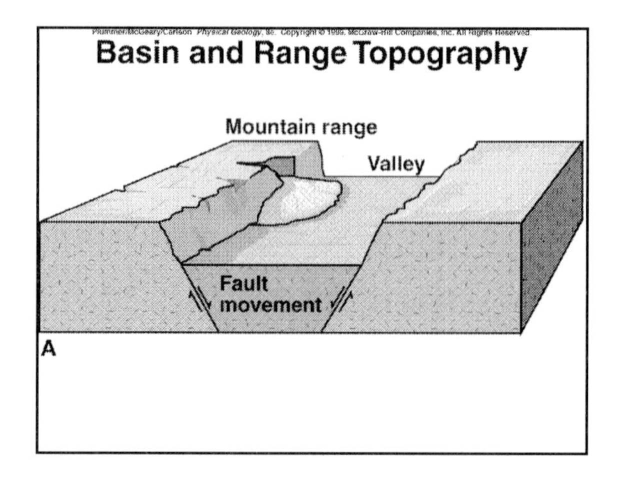
Basin and Range Topography
Mountain range
Valley
Fault
movement
A

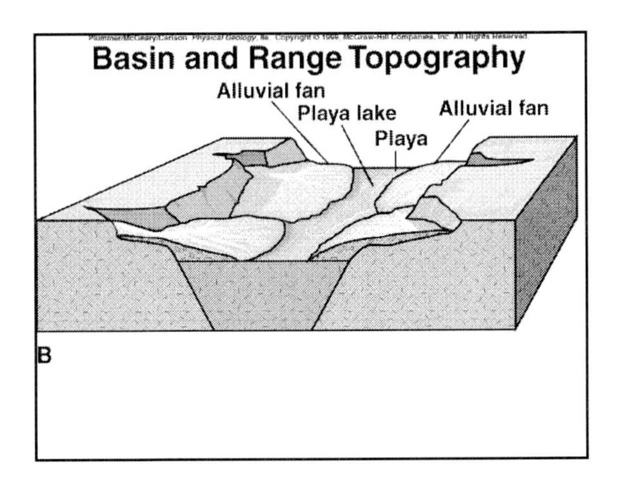
Plummer/McGeary/Carlson Physical Geology, 8e Copyright © 1999. McGraw-Hill Companies, Inc. All Rights Reserved
Basin and Range Topography
Alluvial fan
Playa lake
Alluvial fan
Playa
B

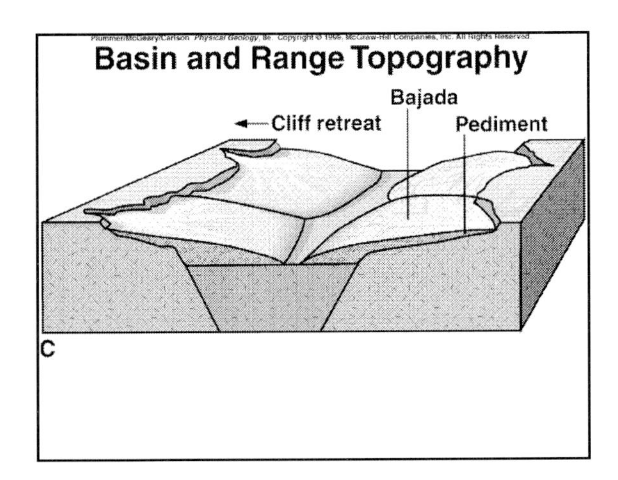
Plummer/McGeary/Carlson Physical Geology, 8e Copyright © 1999. McGraw-Hill Companies, Inc. All Rights Reserved
Basin and Range Topography
Bajada
Cliff retreat
Pediment
C

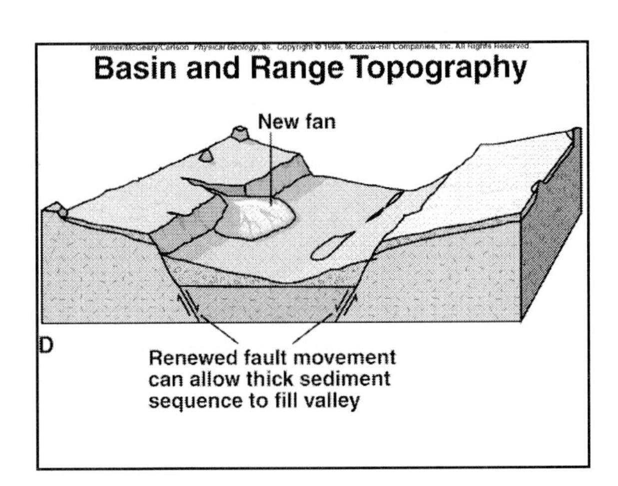
Plummer/McGeary/Carlson Physical Geology, 8e Copyright © 1999. McGraw-Hill Companies, Inc. All Rights Reserved
Basin and Range Topography
New fan
D
Renewed fault movement can allow thick sediment sequence to fill valley

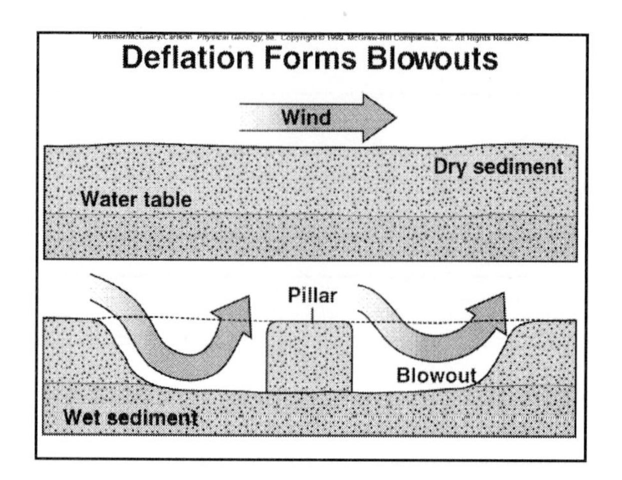
Deflation Forms Blowouts
Wind
Dry sediment
Water table
Pillar
Blowout
Wet sediment

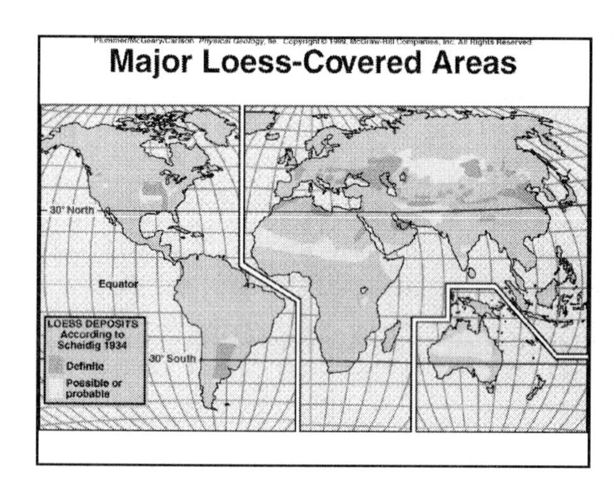
Major Loess-Covered Areas
30° North
Equator
30° South
LOESS DEPOSITS
According to Scheidig 1934
Definite
Possible or probable

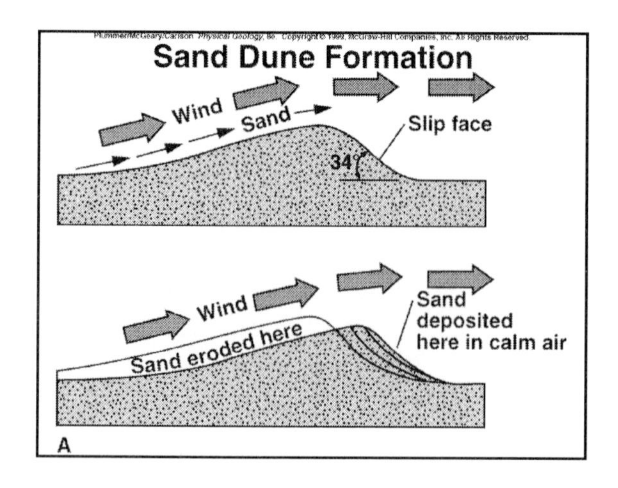
Sand Dune Formation
Wind
Sand
Slip face
34°
Wind
Sand eroded here
Sand deposited here in calm air
A

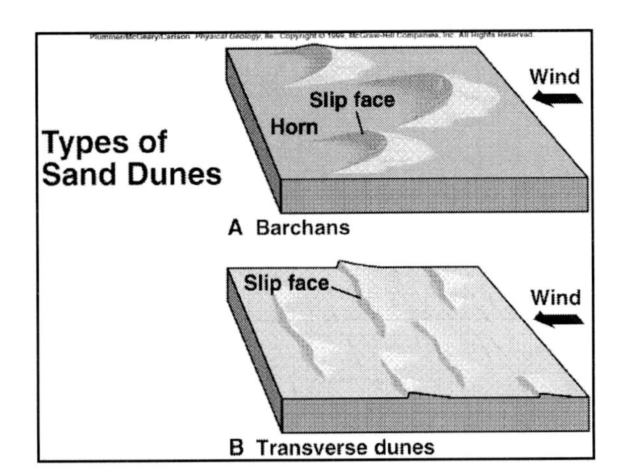
Types of
Sand Dunes
Wind
Slip face
Horn
A Barchans
Slip face
Wind
B Transverse dunes

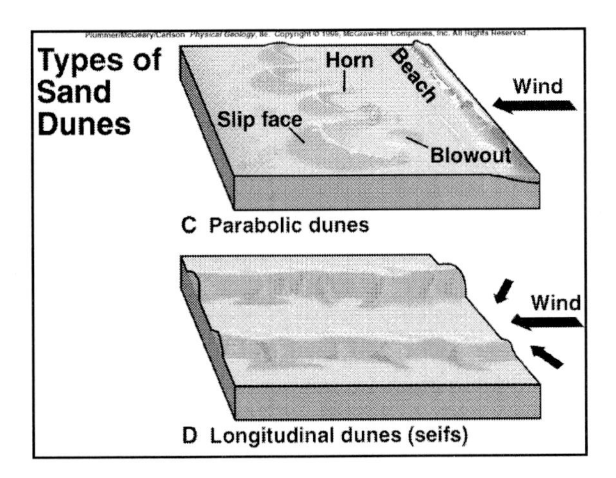
Types of
Sand
Dunes
Horn
Beach
Wind
Slip face
Blowout
C Parabolic dunes
Wind
D Longitudinal dunes (seifs)

Chapter 14

Physical Geology

eighth edition

Plummer/McGeary/Carlson

Waves, Beaches, and Coasts

Water Waves

- Seas, Swells, Surf
- Wave Height
- Crest
- Trough
- Wavelength
- Surf
  - Breaker

Nearshore Circulation

- Wave Refraction
- Longshore Currents
- Rip Currents
- Beaches
  - Beach Face
  - Marine Terrace
    - Wave-built
    - Wave-cut
  - Berm
  - Beach sediment

Longshore Drift of Sediment

- Longshore Drift
  - Swash/Backwash
- Spit
- Baymouth Bar
- Tombolo
- Human Interference with Sand Drift
  - Jetties
  - Groins
  - Breakwater
- Sources of Sand on Beaches

Coasts and Coastal Features

- Erosional Coasts
  - Headlands
  - Coastal Straightening
  - Sea Cliffs
  - Wave-cut Platform
  - Stacks
  - Arches
- Depositional Coasts
  - Barrier Islands
    - Tidal Deltas
  - Deltas
  - Glacial Deposition

Coasts and Coastal Features

- Drowned Coasts
  - Estuaries
  - Fiords
- Uplifted Coasts (Emergent)
  - Uplifted Marine Terraces
- Coasts Shaped by Organisms
  - Algal Reefs
  - Branching Mangrove Roots

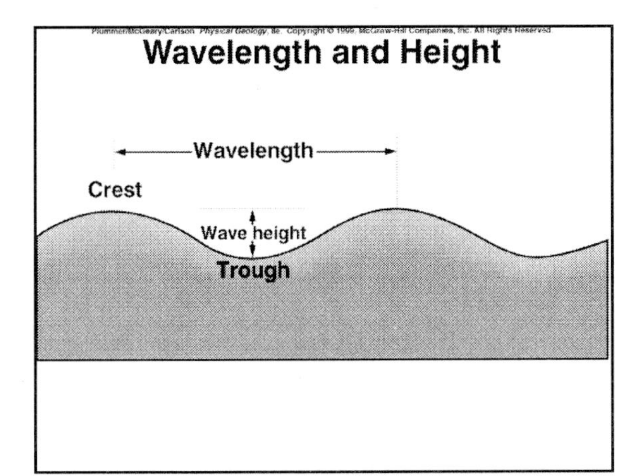

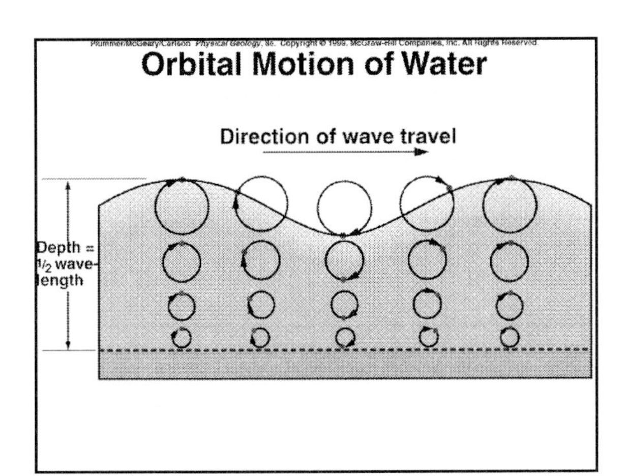

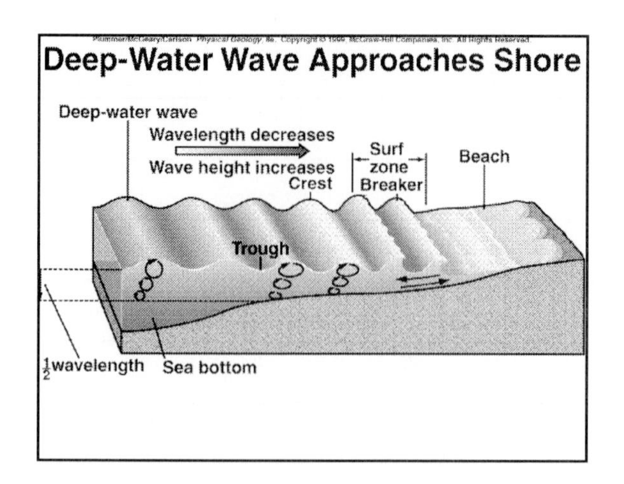
Deep-Water Wave Approaches Shore
Deep-water wave
Wavelength decreases
Wave height increases
Crest
Surf zone
Breaker
Beach
Trough
½ wavelength
Sea bottom

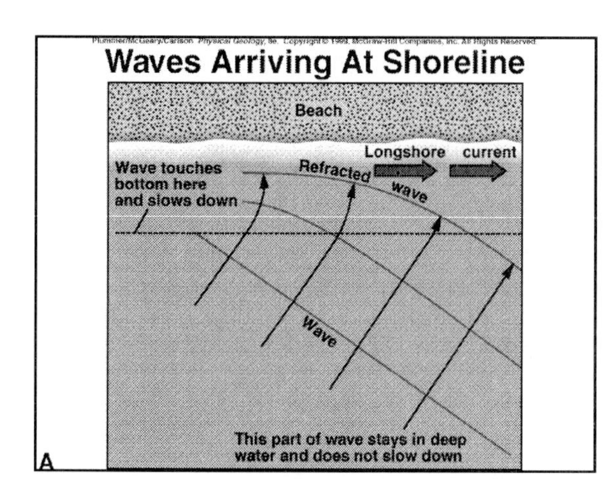
Waves Arriving At Shoreline
Beach
Longshore current
Wave touches bottom here and slows down
Refracted wave
Wave
This part of wave stays in deep water and does not slow down
A

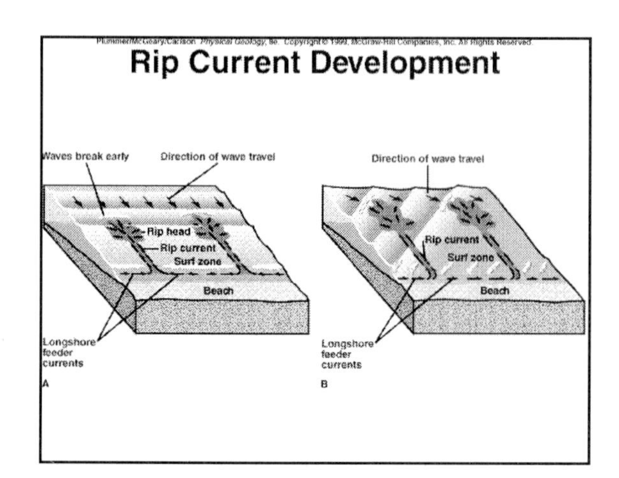
Rip Current Development
Waves break early
Direction of wave travel
Rip head
Rip current
Surf zone
Beach
Longshore feeder currents
A
Direction of wave travel
Rip current
Surf zone
Beach
Longshore feeder currents
B

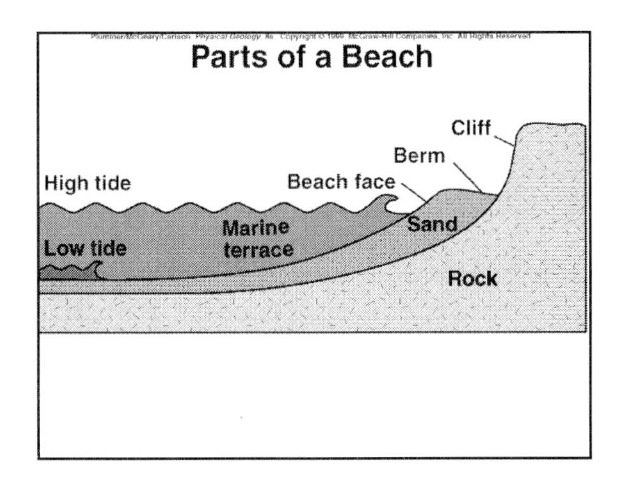
Plummer/McGeary/Carlson Physical Geology, 8e Copyright © 1999 McGraw-Hill Companies, Inc. All Rights Reserved
Parts of a Beach
Cliff
Berm
High tide
Beach face
Marine terrace
Sand
Low tide
Rock

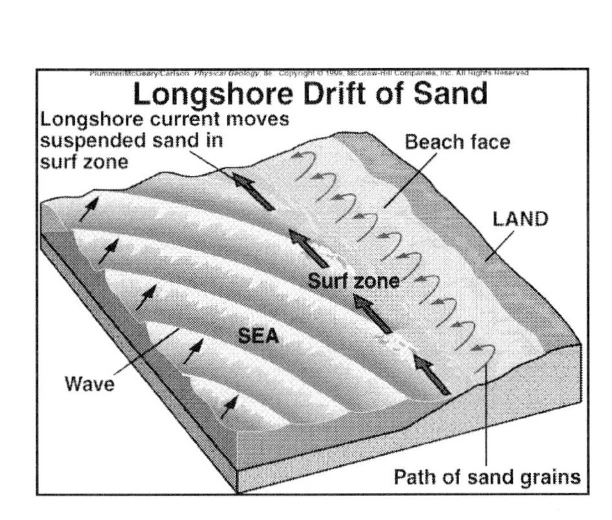
Plummer/McGeary/Carlson Physical Geology, 8e Copyright © 1999 McGraw-Hill Companies, Inc. All Rights Reserved
Longshore Drift of Sand
Longshore current moves suspended sand in surf zone
Beach face
LAND
Surf zone
SEA
Wave
Path of sand grains

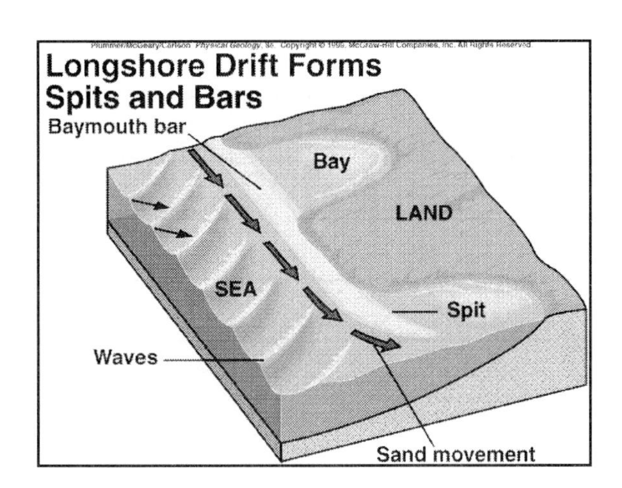
Plummer/McGeary/Carlson Physical Geology, 8e Copyright © 1999 McGraw-Hill Companies, Inc. All Rights Reserved
Longshore Drift Forms Spits and Bars
Baymouth bar
Bay
LAND
SEA
Spit
Waves
Sand movement

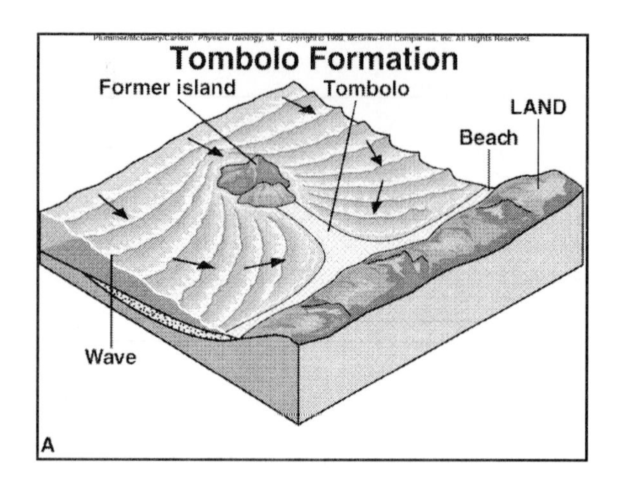
Plummer/McGeary/Carlson Physical Geology, 8e Copyright © 1999. McGraw-Hill Companies, Inc. All Rights Reserved.
Tombolo Formation
Former island
Tombolo
LAND
Beach
Wave
A

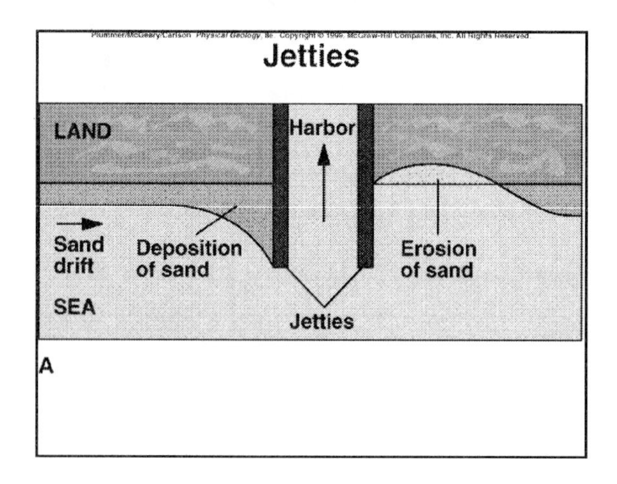
Plummer/McGeary/Carlson Physical Geology, 8e Copyright © 1999. McGraw-Hill Companies, Inc. All Rights Reserved.
Jetties
LAND
Harbor
Sand drift
Deposition of sand
Erosion of sand
SEA
Jetties
A

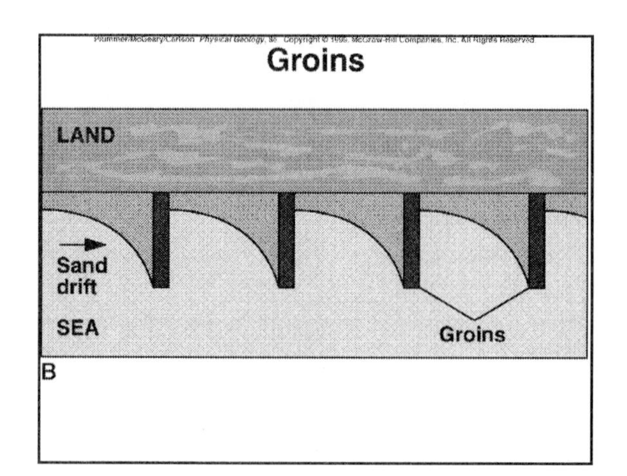
Plummer/McGeary/Carlson Physical Geology, 8e Copyright © 1999. McGraw-Hill Companies, Inc. All Rights Reserved.
Groins
LAND
Sand drift
SEA
Groins
B

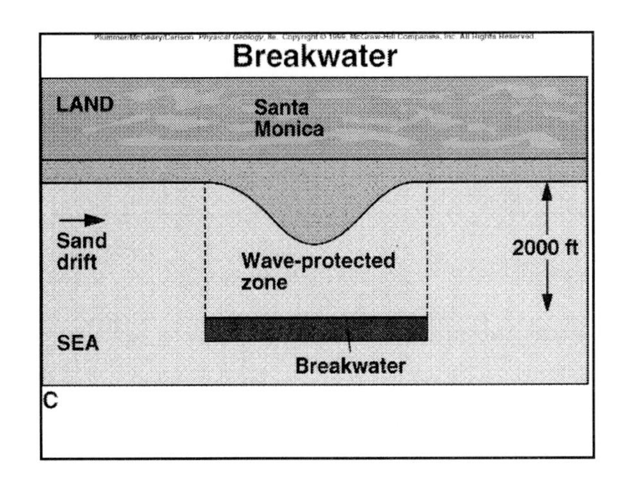
Plummer/McGeary/Carlson Physical Geology, 8e. Copyright © 1999, McGraw-Hill Companies, Inc. All Rights Reserved
Breakwater
LAND
Santa
Monica
Sand
drift
Wave-protected
zone
2000 ft
SEA
Breakwater
C

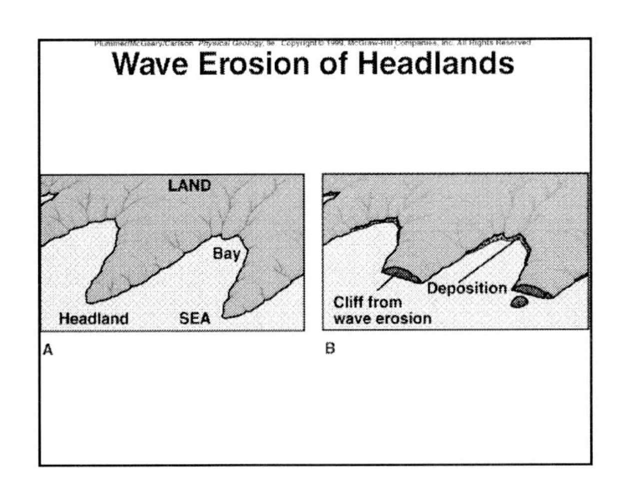
Wave Erosion of Headlands
LAND
Bay
Headland
SEA
A
Deposition
Cliff from
wave erosion
B

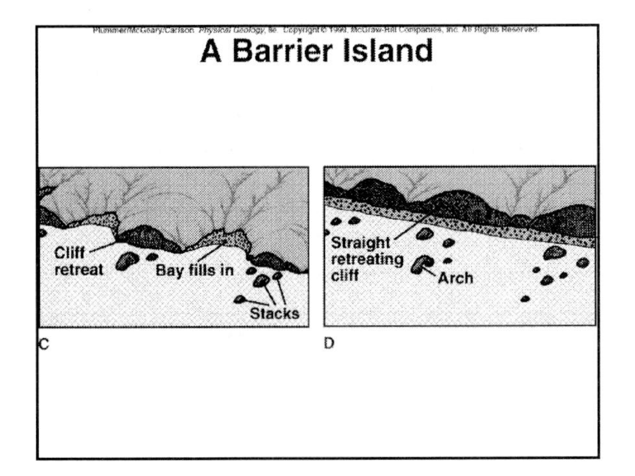
A Barrier Island
Cliff
retreat
Bay fills in
Stacks
C
Straight
retreating
cliff
Arch
D

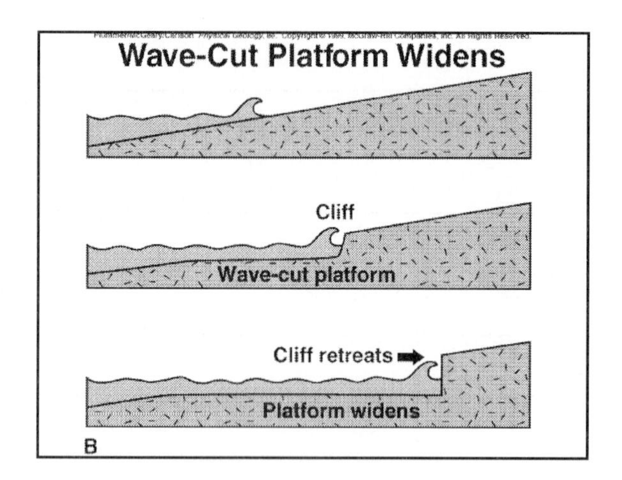
Wave-Cut Platform Widens
Cliff
Wave-cut platform
Cliff retreats
Platform widens
B

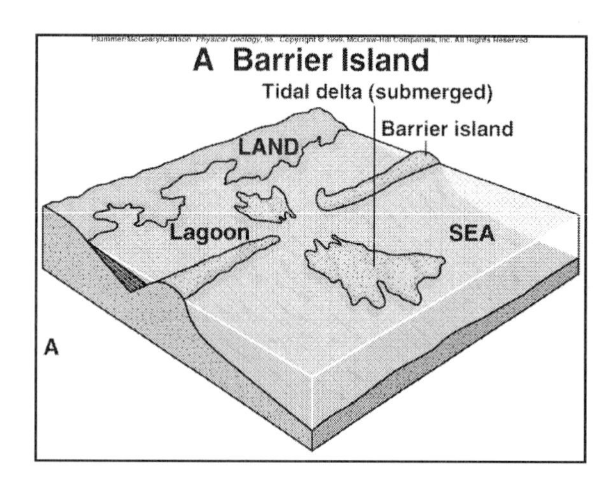
A Barrier Island
Tidal delta (submerged)
Barrier island
LAND
Lagoon
SEA
A

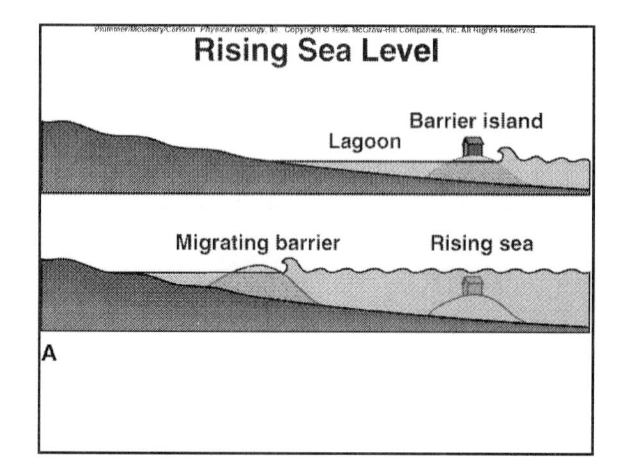
Rising Sea Level
Barrier island
Lagoon
Migrating barrier
Rising sea
A

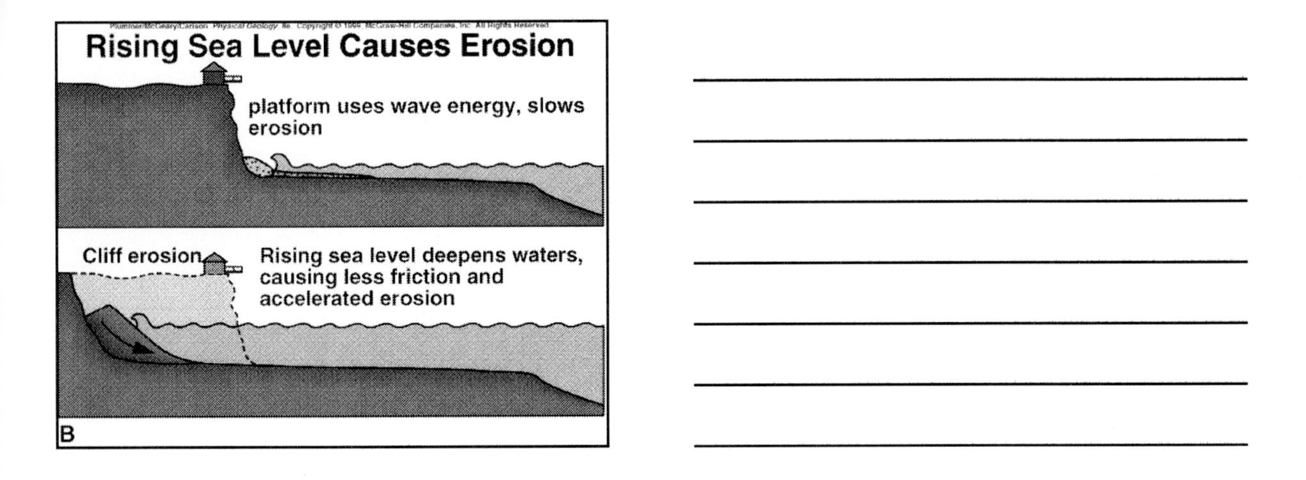
Rising Sea Level Causes Erosion
platform uses wave energy, slows erosion
Cliff erosion
Rising sea level deepens waters, causing less friction and accelerated erosion
B

Chapter 15

Physical Geology

eighth edition

Plummer/McGeary/Carlson

GEOLOGIC STRUCTURES

"Architecture of bedrock"

Structural Geology- concerned with shapes, arrangement, interrelationships of bedrock units & forces that cause them.

TECTONIC FORCES AT WORK

- Stess & Strain
  - Stress
  - Strain
  - Compressive stress
    - Shortening strain
  - Tensional stress
    - stretching or extensional strain
  - Shear Stress
    - Shear strain

Behavior of rocks to stress & strain

- Elastic
  - Elastic limit
- Plastic
- Brittle

Present Deformation of the Crust

- Active ***fault***
- Other movement of the crust

Structures as a Record of the Geologic Past

- Geologic maps and field methods
  - Observations of *outcrops*
  - Geologic map
  - Strike and dip
    - *Original horizontality*
    - *Strike*
    - *Angle of dip*
    - *Direction of dip*
    - Symbols
  - Geologic cross section

FOLDS

- Anticline vs. syncline
  - Hinge line (axis)
  - Limb
  - Axial plane
- Plunging fold
- Structural dome
- Structural basin

FOLDS

- Interpreting folds
  - Open fold
  - Isoclinal fold
  - Overturned fold
  - Recumbent fold

Fractures in rock

- Joints
  - Columnar jointing
  - Sheet jointing
  - Joint set
- Faults
  - Dip-slip faults- normal and reverse
    - Footwall vs. hanging wall
  - *Normal fault*
    - Graben; Rift
    - Horst ; fault-block mountain range

Fractures in rock

- Faults
  - *Reverse fault*
    - Thrust fault- low angle reverse fault
  - *Strike-Slip fault*
    - Left-lateral vs. right-lateral

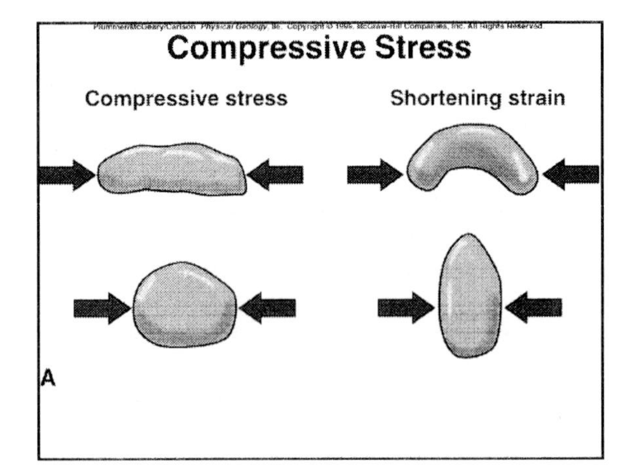

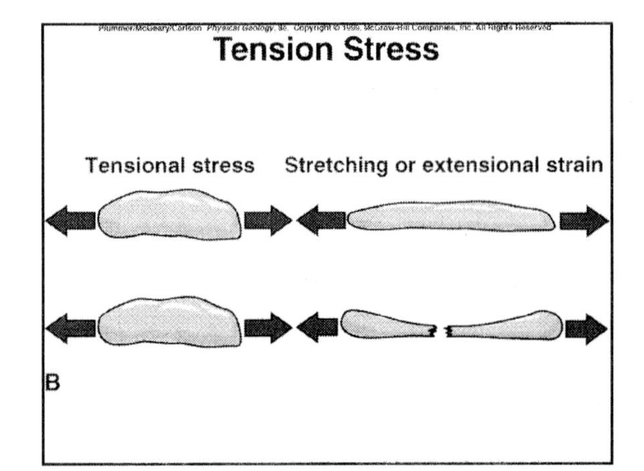

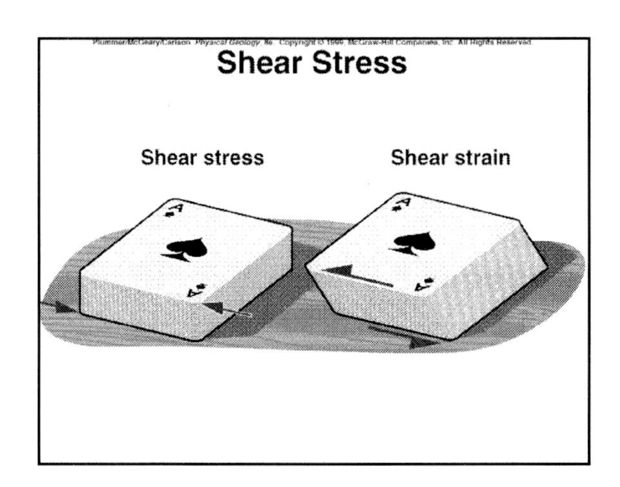
Shear Stress
Shear stress
Shear strain

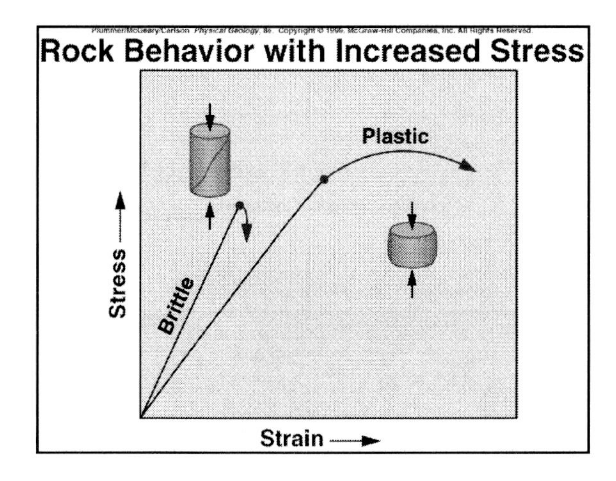
Rock Behavior with Increased Stress
Plastic
Brittle
Stress
Strain

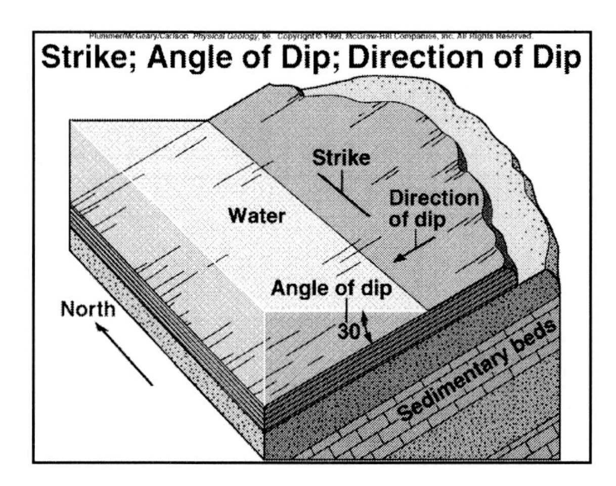
Strike; Angle of Dip; Direction of Dip
Strike
Direction of dip
Water
Angle of dip
30
North
Sedimentary beds

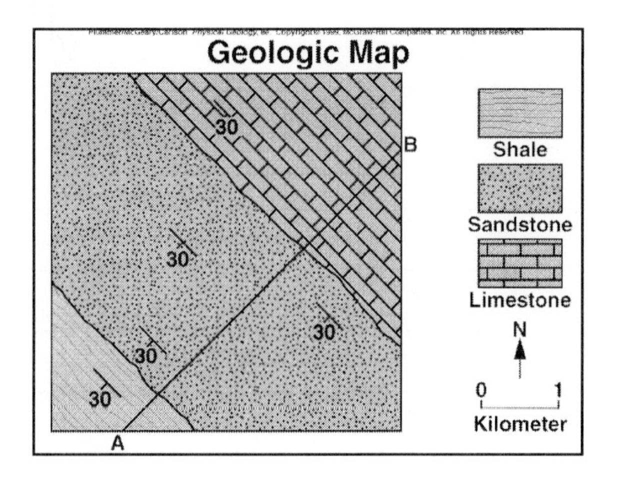
Plummer/McGeary/Carlson Physical Geology, 8e Copyright © 1999 McGraw-Hill Companies, Inc. All Rights Reserved.
Geologic Map
30
B
30
30
30
30
A
Shale
Sandstone
Limestone
N
0
1
Kilometer

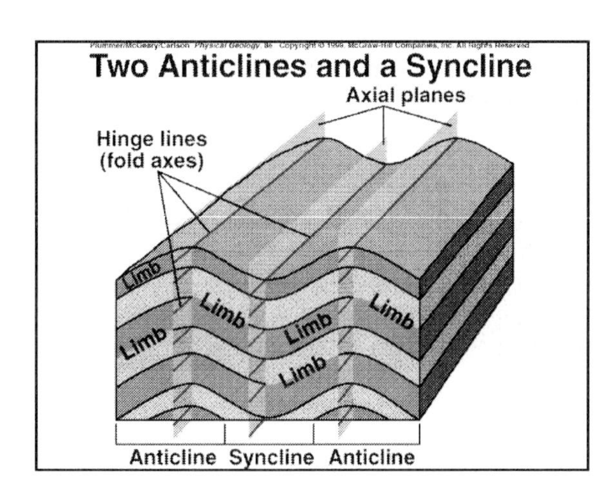
Plummer/McGeary/Carlson Physical Geology, 8e Copyright © 1999 McGraw-Hill Companies, Inc. All Rights Reserved.
Two Anticlines and a Syncline
Axial planes
Hinge lines
(fold axes)
Limb
Limb
Limb
Limb
Limb
Limb
Anticline
Syncline
Anticline

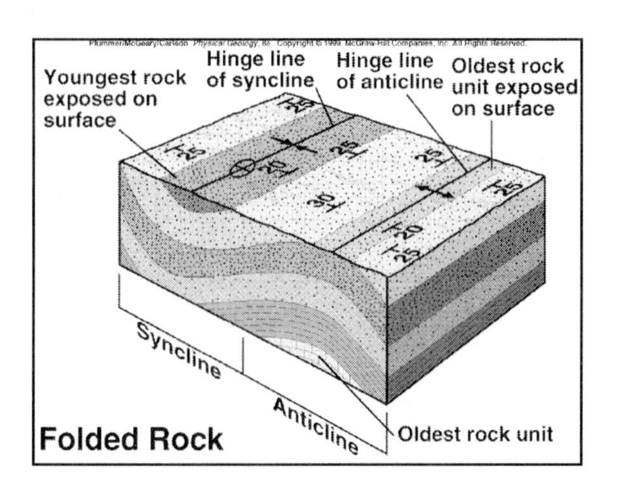
Plummer/McGeary/Carlson Physical Geology, 8e Copyright © 1999 McGraw-Hill Companies, Inc. All Rights Reserved.
Youngest rock
exposed on
surface
Hinge line
of syncline
Hinge line
of anticline
Oldest rock
unit exposed
on surface
Syncline
Anticline
Oldest rock unit
Folded Rock

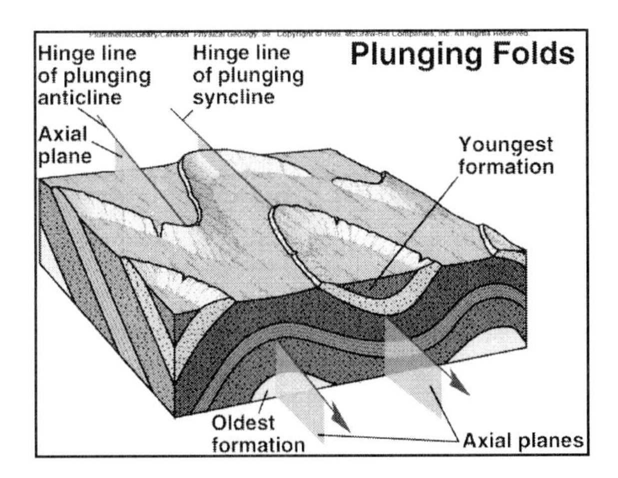
Plunging Folds
Hinge line
of plunging
anticline
Hinge line
of plunging
syncline
Axial
plane
Youngest
formation
Oldest
formation
Axial planes

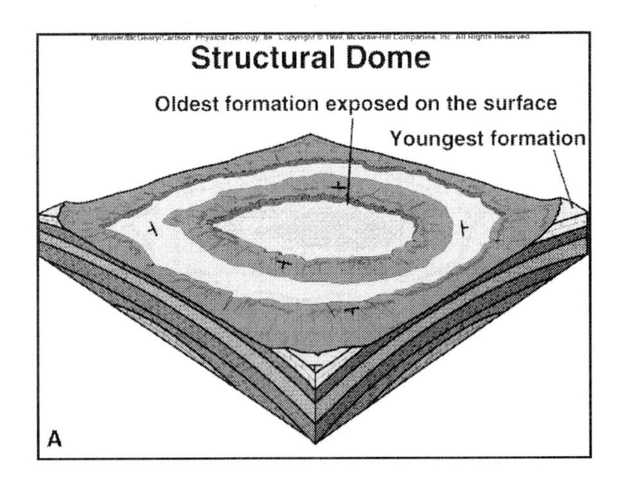
Structural Dome
Oldest formation exposed on the surface
Youngest formation
A

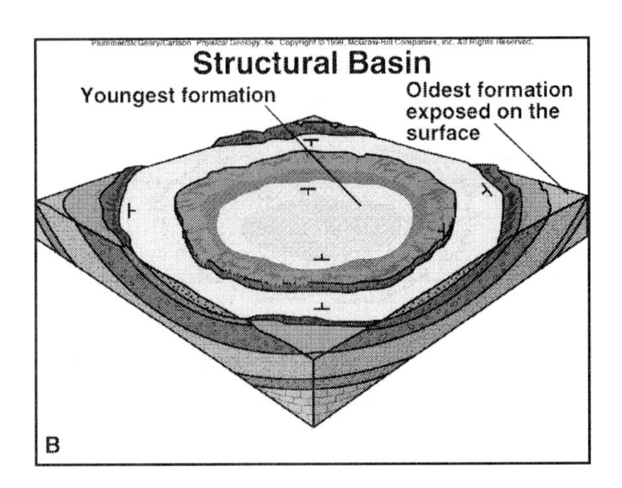
Structural Basin
Youngest formation
Oldest formation
exposed on the
surface
B

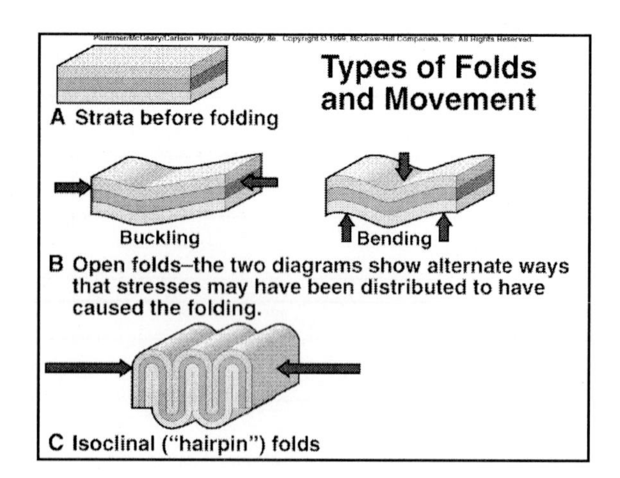
Plummer/McGeary/Carlson Physical Geology, 8e Copyright © 1999, McGraw-Hill Companies, Inc. All Rights Reserved
Types of Folds and Movement
A Strata before folding
Buckling
Bending
B Open folds–the two diagrams show alternate ways that stresses may have been distributed to have caused the folding.
C Isoclinal ("hairpin") folds

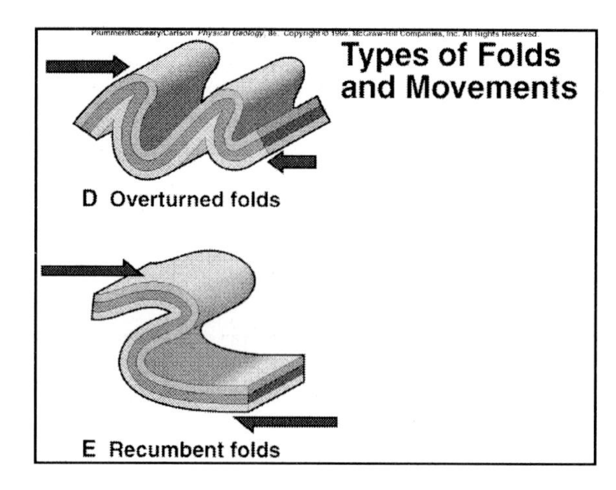
Plummer/McGeary/Carlson Physical Geology, 8e Copyright © 1999, McGraw-Hill Companies, Inc. All Rights Reserved
Types of Folds and Movements
D Overturned folds
E Recumbent folds

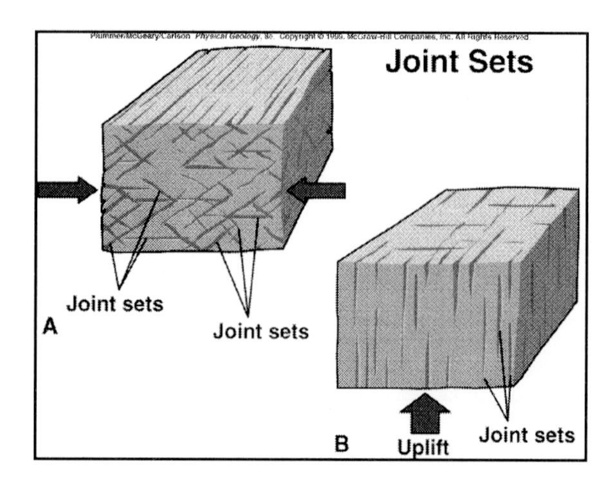
Plummer/McGeary/Carlson Physical Geology, 8e Copyright © 1999, McGraw-Hill Companies, Inc. All Rights Reserved
Joint Sets
Joint sets
A
Joint sets
B
Uplift
Joint sets

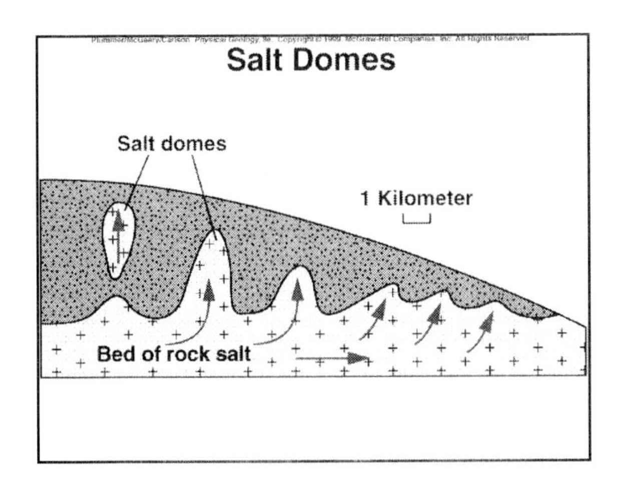
Salt Domes
Salt domes
1 Kilometer
Bed of rock salt

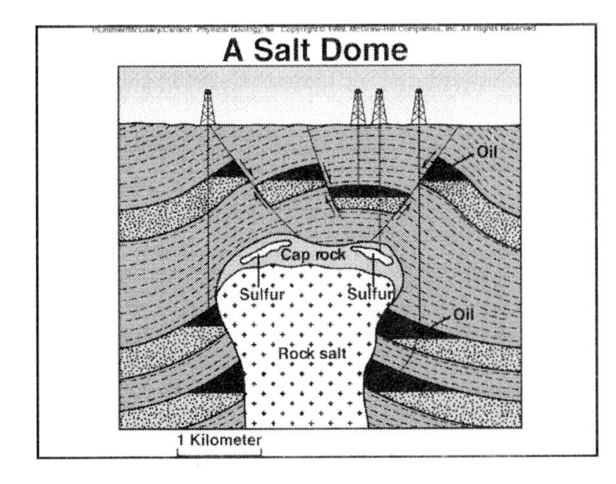
A Salt Dome
Oil
Cap rock
Sulfur
Sulfur
Oil
Rock salt
1 Kilometer

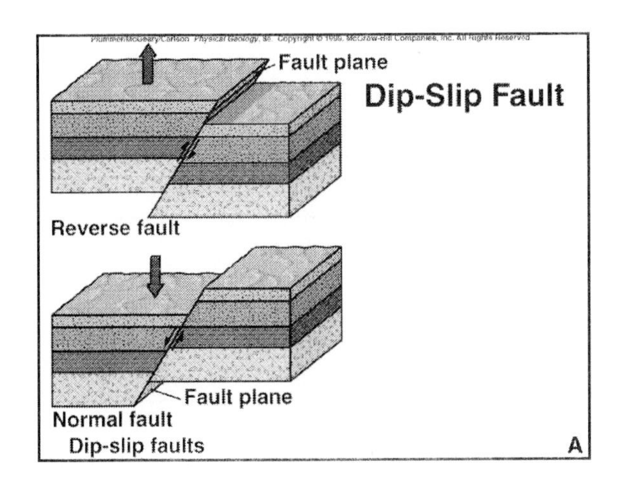
Fault plane
Dip-Slip Fault
Reverse fault
Fault plane
Normal fault
Dip-slip faults
A

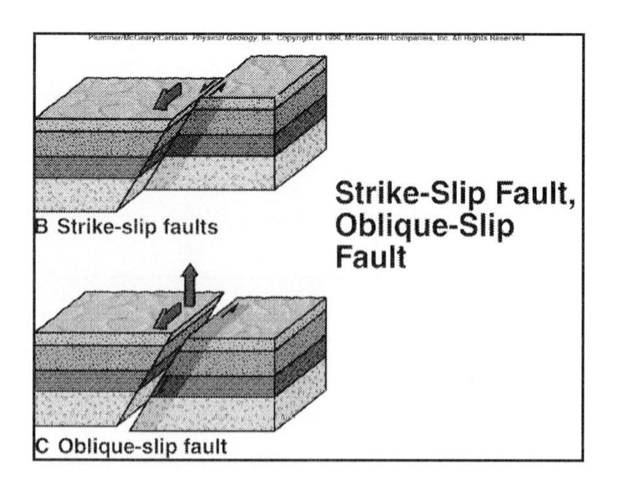
Plummer/McGeary/Carlson Physical Geology, 8e. Copyright © 1999, McGraw-Hill Companies, Inc. All Rights Reserved.
Strike-Slip Fault, Oblique-Slip Fault
B Strike-slip faults
C Oblique-slip fault

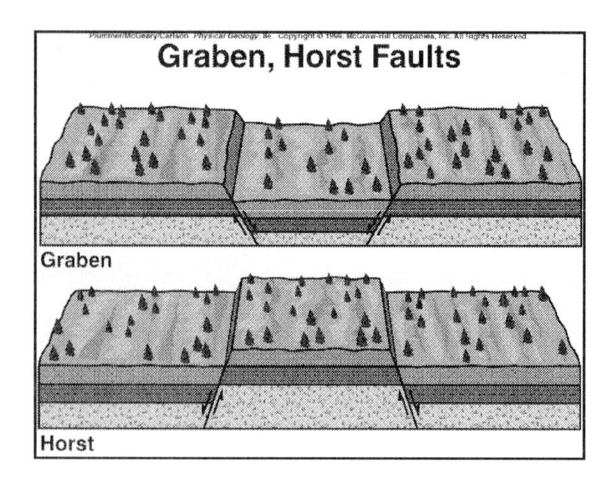
Graben, Horst Faults
Graben
Horst

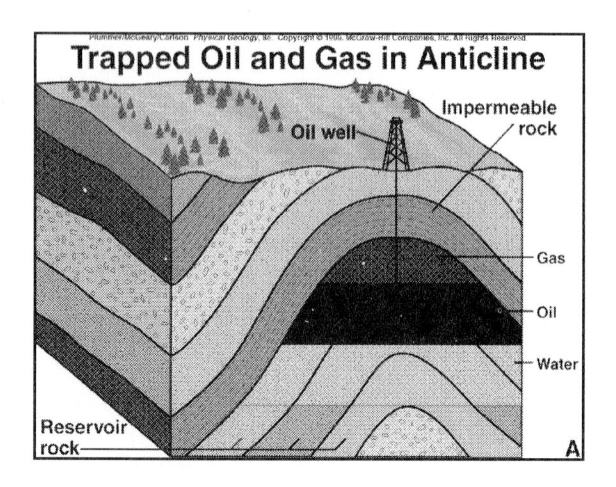
Trapped Oil and Gas in Anticline
Oil well
Impermeable rock
Gas
Oil
Water
Reservoir rock
A

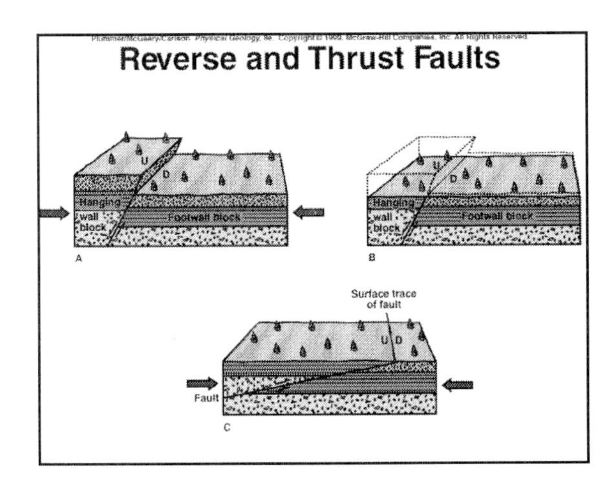
Reverse and Thrust Faults
Hanging wall block
Footwall block
U
D
A
B
Surface trace of fault
Fault
C

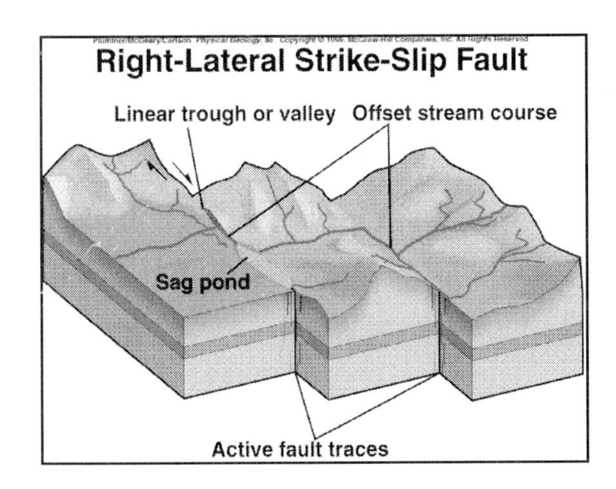
Right-Lateral Strike-Slip Fault
Linear trough or valley
Offset stream course
Sag pond
Active fault traces

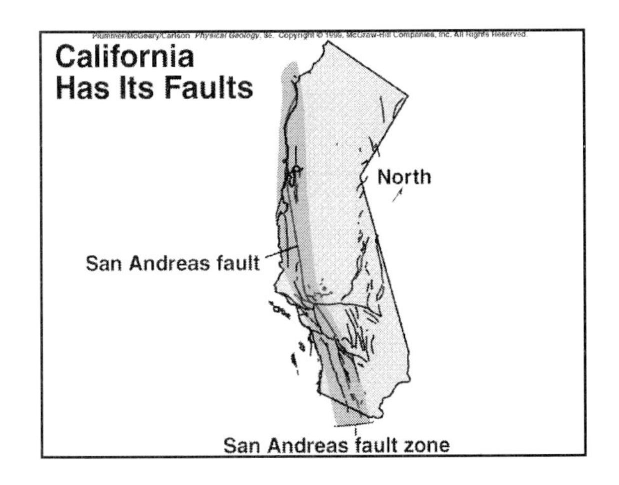
California Has Its Faults
North
San Andreas fault
San Andreas fault zone

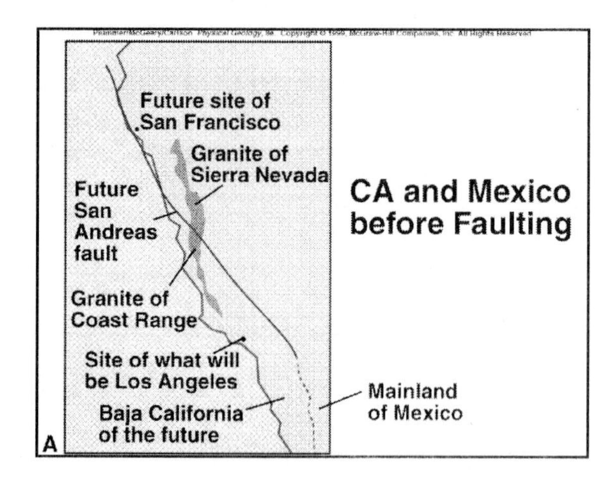
Future site of San Francisco
Granite of Sierra Nevada
Future San Andreas fault
Granite of Coast Range
Site of what will be Los Angeles
Baja California of the future
Mainland of Mexico
A
CA and Mexico before Faulting

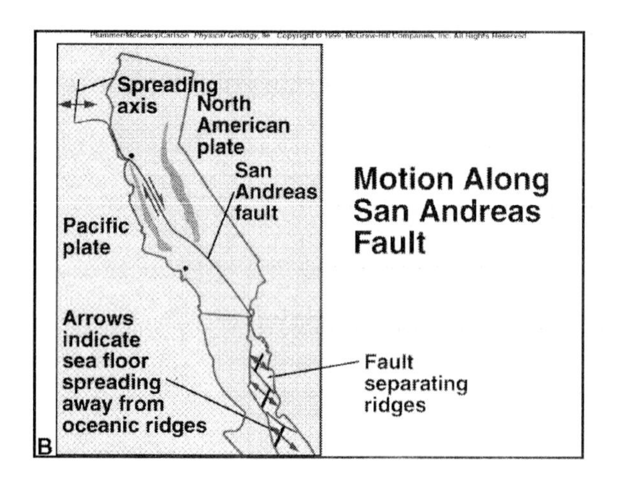
Spreading axis
North American plate
San Andreas fault
Pacific plate
Arrows indicate sea floor spreading away from oceanic ridges
Fault separating ridges
B
Motion Along San Andreas Fault

Chapter 16

Physical Geology

eighth edition

Plummer/McGeary/Carlson

EARTHQUAKES

Causes

- Sudden release of energy stored in rocks
  - Released as seismic waves
- Elastic Rebound Theory
  - Fault motion
  - Tectonic forces
- Also associated with volcanic activity

Seismic Waves

- Focus
- Epicenter
- Body waves
  - **P wave**
    - can pass through solids and fluids
  - **S wave**
    - slower
    - can pass through solids only
- Surface waves- most damaging
  - Love waves
  - Rayleigh waves

Locating earthquakes

- Seismometer
  - Seismograph; seismogram
- Determining location
  - Travel-time curves
  - Depth of focus

Measuring size of an earthquake

- Intensity
  - Modified Mercalli Scale
    - I to XII
- Magnitude
  - Richter Scale
  - Moment magnitude
- Location & size of earthquakes in U.S.

Effects of earthquakes

- Ground motion
- Fire
- Landslides
- Liquifaction
- Permanent displacement of land surface
- Aftershocks
- Tsunamis- seismic sea waves

Earthquake Distribution

- Circum-Pacific Belt
- Mediterranean-Himalayan Belt
- Mid-oceanic ridge
- Benioff zones
  - Extend from trenches beneath continents or island arcs

First-motion studies of earthquakes

- Determining push or pull from seismograms
- Need to know orientation of fault trace

Earthquakes and plate tectonics

- Earthquakes at plate boundaries
  - Divergent boundaries
    - Along sides of rift valley
    - Indicate horizontal extension
  - Transform boundaries
    - Shallow focus
  - Convergent boundaries
    - Collision boundaries between continents
    - Subduction
- Subduction angle

Earthquakes prediction

- Scientific techniques being explored
  - Microseisms
  - Properties of the rock
  - Water levels in wells
  - Radon emission from wells
  - Surface tilts & changes of elevation
  - Animal behavior
  - Patterns of earthquakes in space & time

Earthquake control

- Water under high pressure
- Release of strain

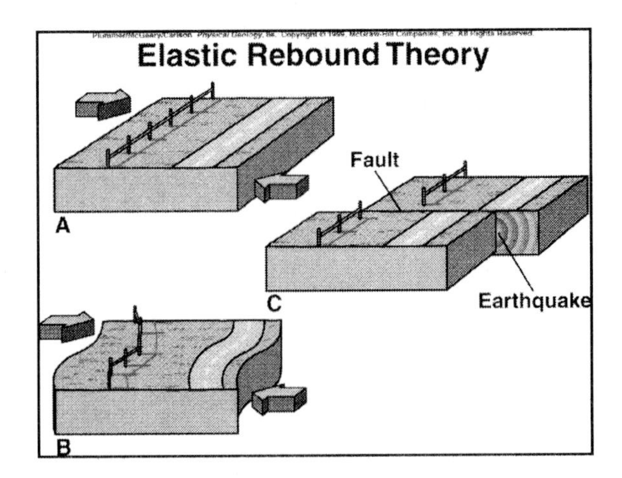
Elastic Rebound Theory
Fault
A
C
Earthquake
B

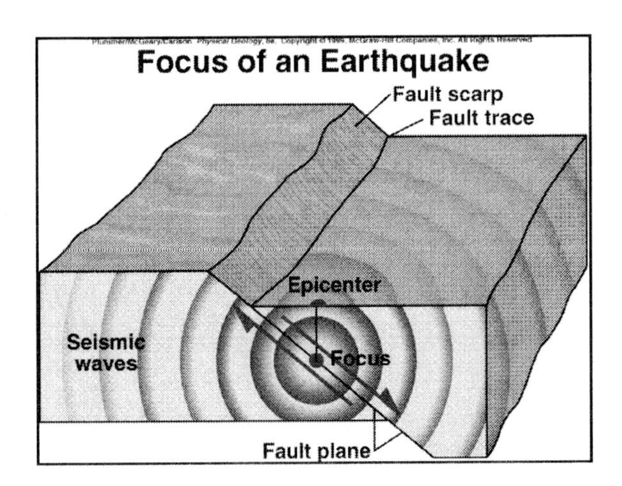
Focus of an Earthquake
Fault scarp
Fault trace
Epicenter
Seismic waves
Focus
Fault plane

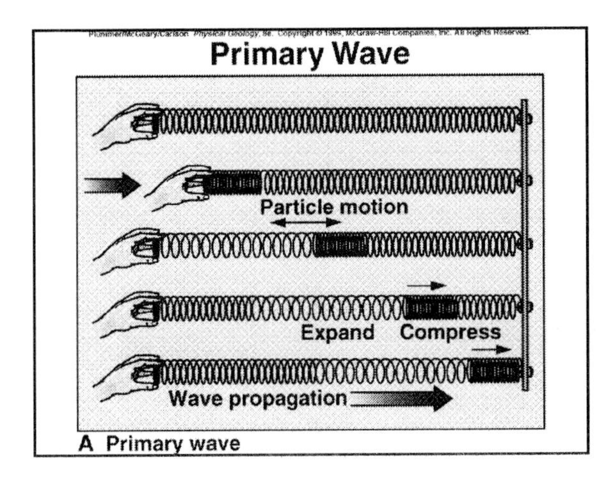
Primary Wave
Particle motion
Expand
Compress
Wave propagation
A Primary wave

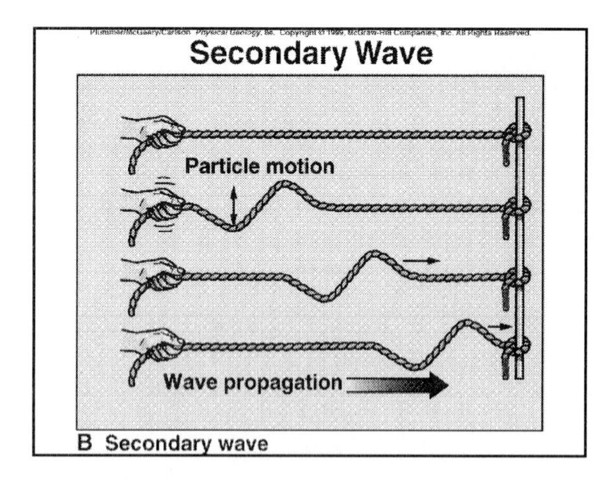
Plummer/McGeary/Carlson Physical Geology, 8e. Copyright © 1999, McGraw-Hill Companies, Inc. All Rights Reserved.
Secondary Wave
Particle motion
Wave propagation
B Secondary wave

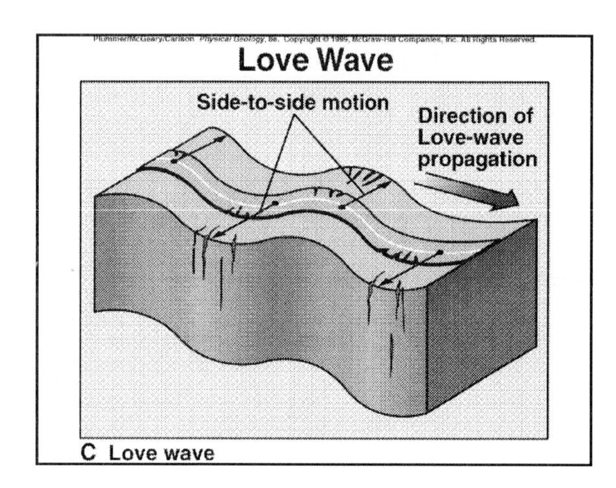
Plummer/McGeary/Carlson Physical Geology, 8e. Copyright © 1999, McGraw-Hill Companies, Inc. All Rights Reserved.
Love Wave
Side-to-side motion
Direction of Love-wave propagation
C Love wave

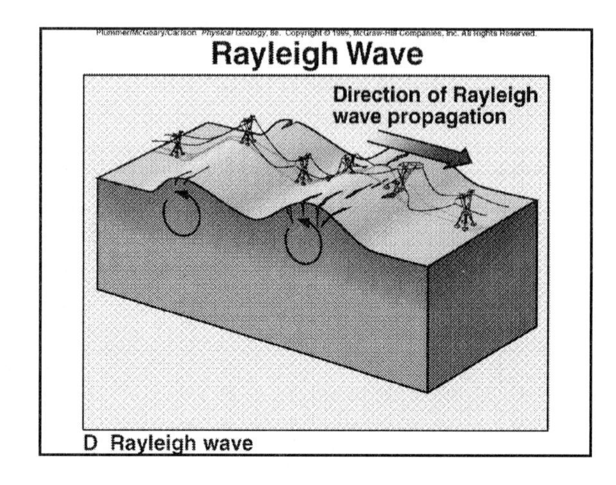
Plummer/McGeary/Carlson Physical Geology, 8e. Copyright © 1999, McGraw-Hill Companies, Inc. All Rights Reserved.
Rayleigh Wave
Direction of Rayleigh wave propagation
D Rayleigh wave

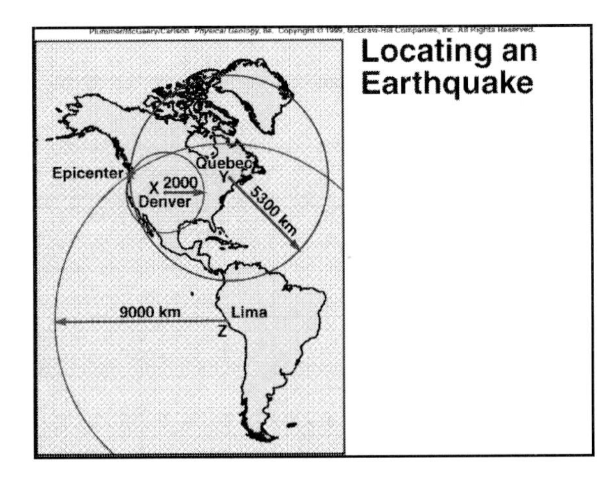
Plummer/McGeary/Carlson Physical Geology, 8e. Copyright © 1999, McGraw-Hill Companies, Inc. All Rights Reserved.
Locating an Earthquake
Epicenter
X 2000
Denver
Quebec
Y
5300 km
9000 km
Lima
Z

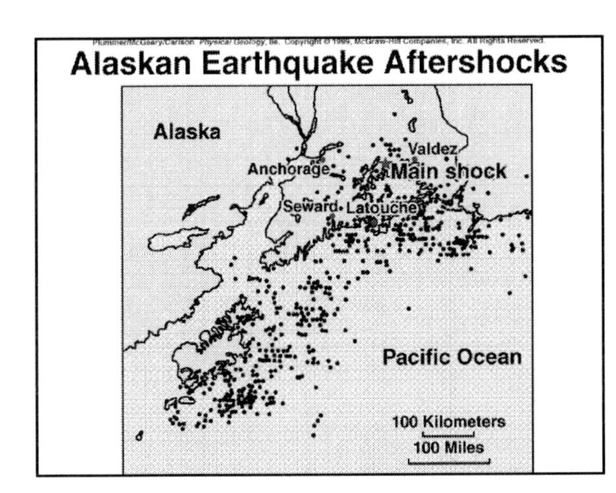
Plummer/McGeary/Carlson Physical Geology, 8e. Copyright © 1999, McGraw-Hill Companies, Inc. All Rights Reserved.
Alaskan Earthquake Aftershocks
Alaska
Valdez
Anchorage
Main shock
Seward
Latouche
Pacific Ocean
100 Kilometers
100 Miles

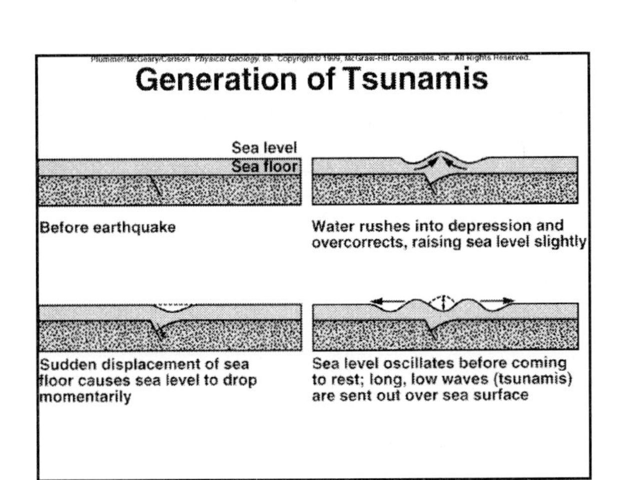
Plummer/McGeary/Carlson Physical Geology, 8e. Copyright © 1999, McGraw-Hill Companies, Inc. All Rights Reserved.
Generation of Tsunamis
Sea level
Sea floor
Before earthquake
Water rushes into depression and overcorrects, raising sea level slightly
Sudden displacement of sea floor causes sea level to drop momentarily
Sea level oscillates before coming to rest; long, low waves (tsunamis) are sent out over sea surface

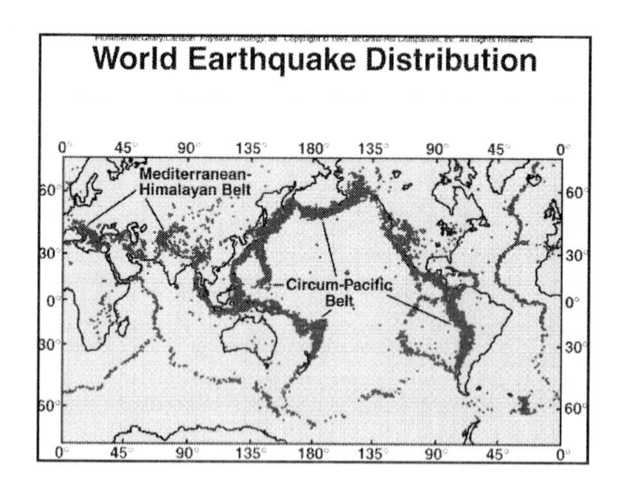
World Earthquake Distribution
Mediterranean-
Himalayan Belt
Circum-Pacific
Belt
0° 45° 90° 135° 180° 135° 90° 45° 0°
60° 30° 0° 30° 60°

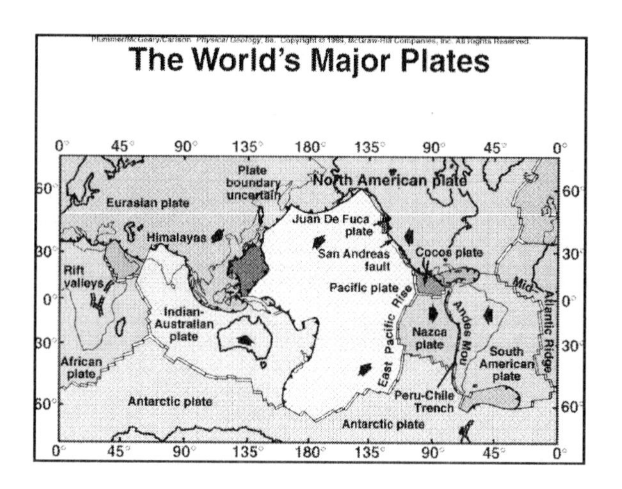
The World's Major Plates
Plate
boundary
uncertain
North American plate
Eurasian plate
Juan De Fuca
plate
Himalayas
San Andreas
fault
Cocos plate
Rift
valleys
Pacific plate
Mid-
Atlantic Ridge
Indian-
Australian
plate
East Pacific Rise
Andes Mou
Nazca
plate
South
American
plate
African
plate
Peru-Chile
Trench
Antarctic plate
Antarctic plate
0° 45° 90° 135° 180° 135° 90° 45° 0°
60° 30° 0° 30° 60°

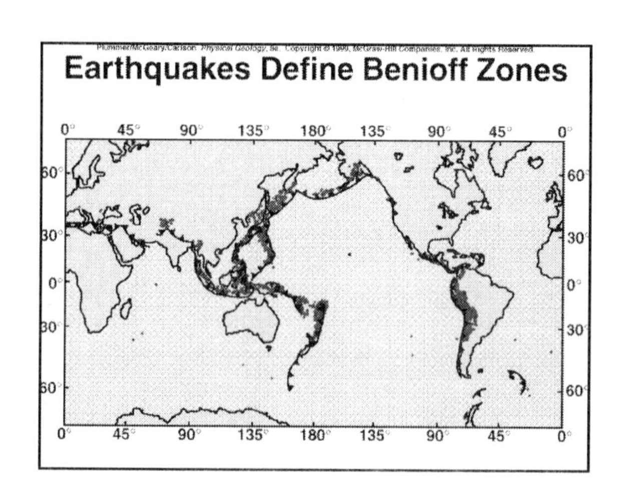
Earthquakes Define Benioff Zones
0° 45° 90° 135° 180° 135° 90° 45° 0°
60° 30° 0° 30° 60°

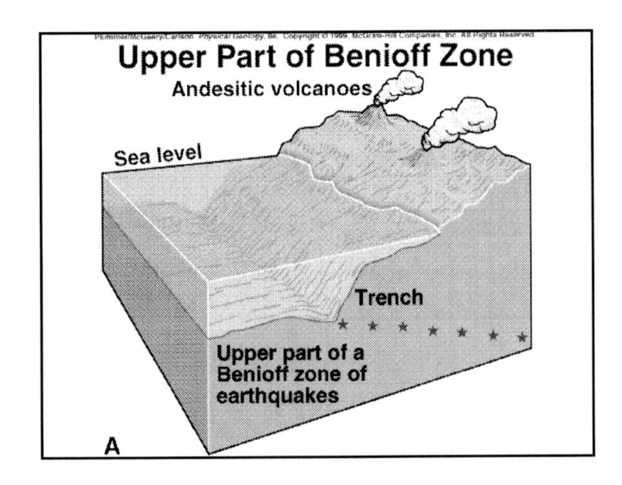
Upper Part of Benioff Zone
Andesitic volcanoes
Sea level
Trench
Upper part of a
Benioff zone of
earthquakes
A

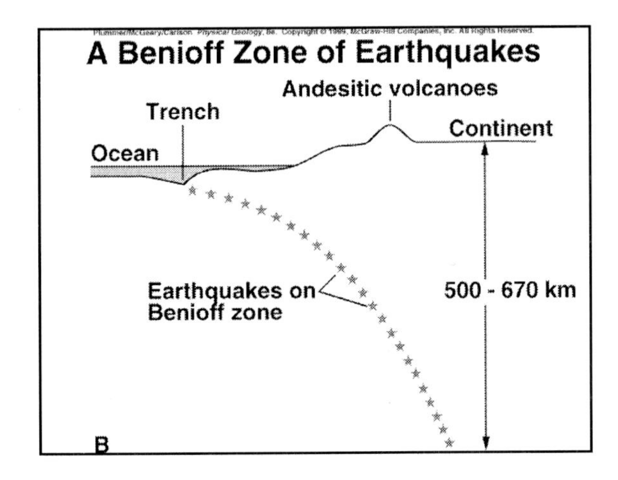
A Benioff Zone of Earthquakes
Andesitic volcanoes
Trench
Continent
Ocean
Earthquakes on
Benioff zone
500 - 670 km
B

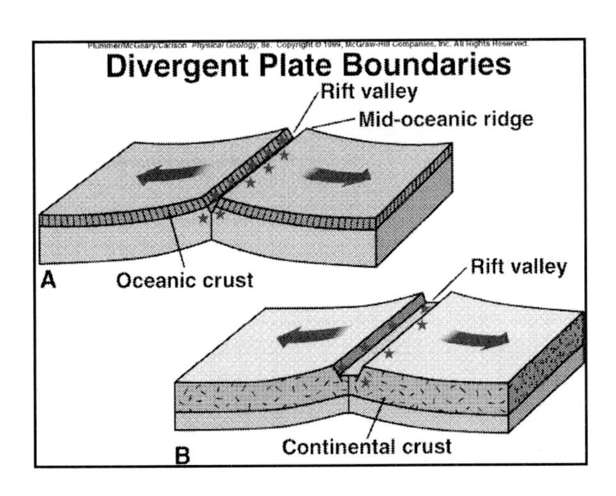
Divergent Plate Boundaries
Rift valley
Mid-oceanic ridge
A
Oceanic crust
Rift valley
B
Continental crust

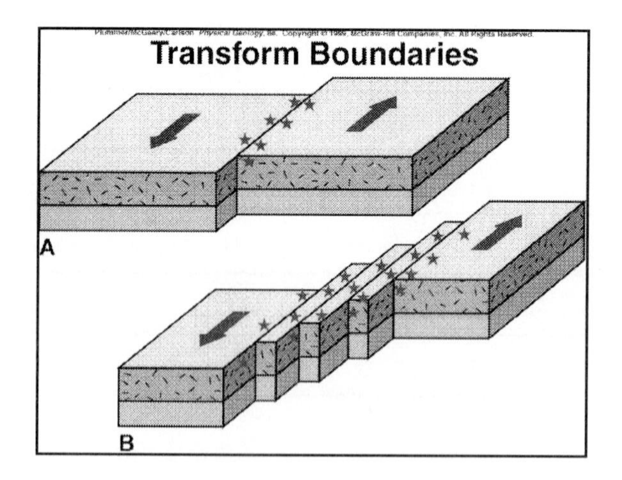
Plummer/McGeary/Carlson Physical Geology, 8e. Copyright © 1999, McGraw-Hill Companies, Inc. All Rights Reserved.
Transform Boundaries
A
B

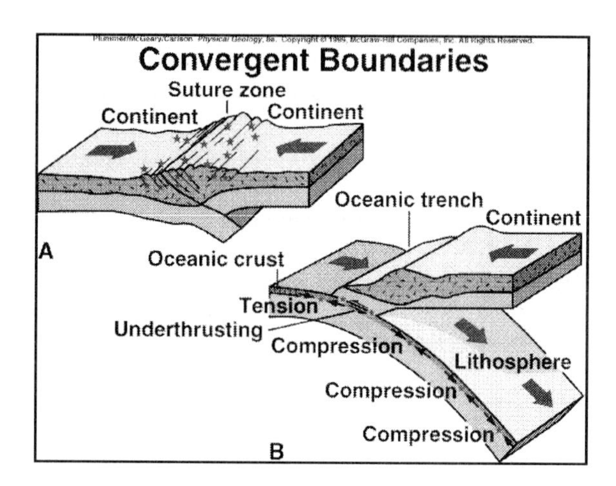
Plummer/McGeary/Carlson Physical Geology, 8e. Copyright © 1999, McGraw-Hill Companies, Inc. All Rights Reserved.
Convergent Boundaries
Suture zone
Continent
Continent
A
Oceanic trench
Continent
Oceanic crust
Tension
Underthrusting
Compression
Lithosphere
Compression
Compression
B

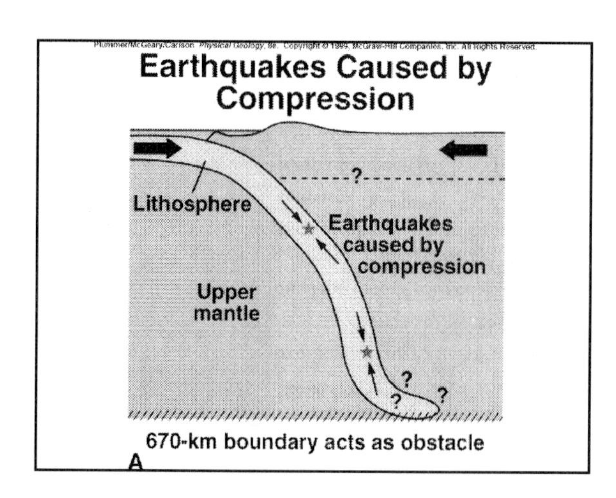
Plummer/McGeary/Carlson Physical Geology, 8e. Copyright © 1999, McGraw-Hill Companies, Inc. All Rights Reserved.
Earthquakes Caused by Compression
Lithosphere
Earthquakes caused by compression
Upper mantle
670-km boundary acts as obstacle
A

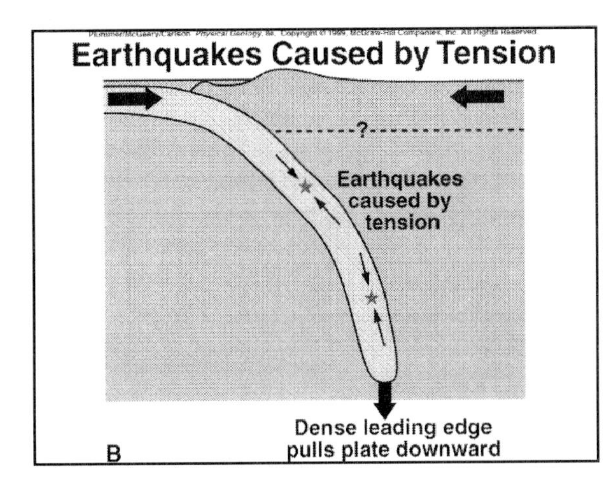
Earthquakes Caused by Tension
?
Earthquakes caused by tension
Dense leading edge pulls plate downward
B

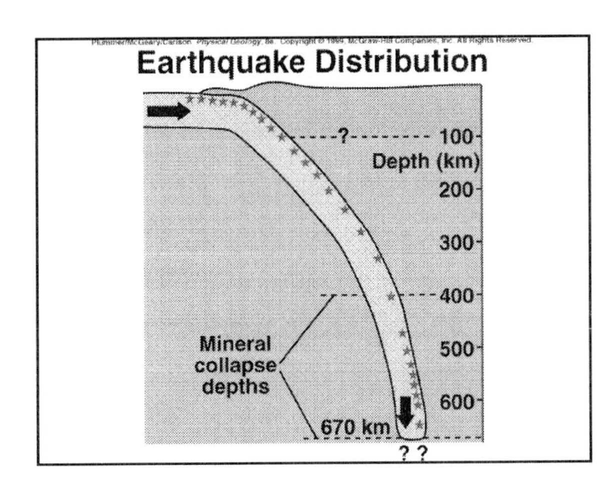
Earthquake Distribution
?
100
Depth (km)
200
300
400
500
600
Mineral collapse depths
670 km
? ?

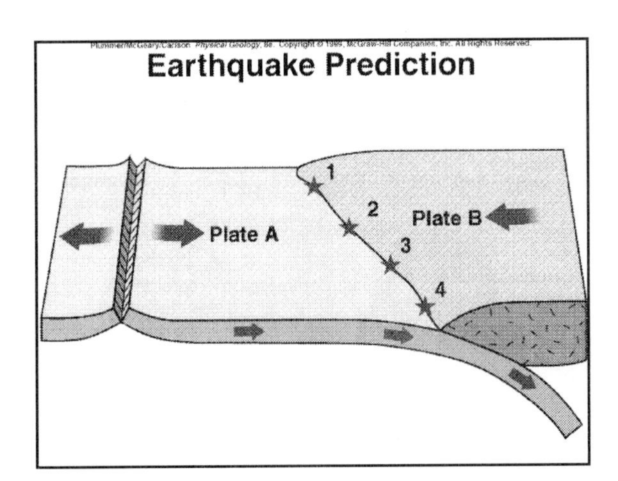
Earthquake Prediction
Plate A
Plate B
1
2
3
4

Chapter 17

Physical Geology

eighth edition

Plummer/McGeary/Carlson

The Earth's Interior

The Earth's Interior

- Most of what we know is from *geophysics*

Evidence from Seismic Waves

- Seismic Reflection
- Seismic Refraction

Earth's Internal Structure

- The Earth's Crust
  - Continental crust- *sial*
    - » Less dense and thicker than oceanic crust
  - Oceanic crust- *sima*
  - Mohorovicic discontinuity

Earth's Internal Structure

- The Mantle
  - Ultramafic Rock
  - Lithosphere
    - » Crust & uppermost mantle
  - Asthenosphere
    - » Low velocity zone
  - Lower mantle

Earth's Internal Structure

- The Core
  - P-wave Shadow Zone
  - S-wave Shadow Zones
  - Liquid Outer Core
  - Solid Inner Core

Earth's Internal Structure

- Composition of the Core
  - Made of metal
- The Core-Mantle Boundary
  - Convection in both core and mantle

Isostasy

- Equilibrium
- Isostatic adjustment
  - Crust "floating" on mantle
- Depth of Equal Pressure
- Crustal Rebound
- Mountain ranges

Gravity Measurements

- Equation for force of gravity
- Measuring gravity
  - Gravity Meter
- Gravity anomalies
  - Positive
  - Negative

Earth's Magnetic Field

- Magnetic Field
- Magnetic Poles
  - Dipolar
  - Source

Earth's Magnetic Field

- Magnetic Reversals
  - Normal & reversed polarity
  - Curie Point
  - Paleomagnetism
    - » Record of magnetic reversals
- Magnetic Anomalies
  - Magnetometer
  - Positive/ Negative Magnetic Anomalies

Heat Within the Earth

- Geothermal Gradient
  - Geothermal Energy
- Temperature in the earth's interior
- Heat Flow

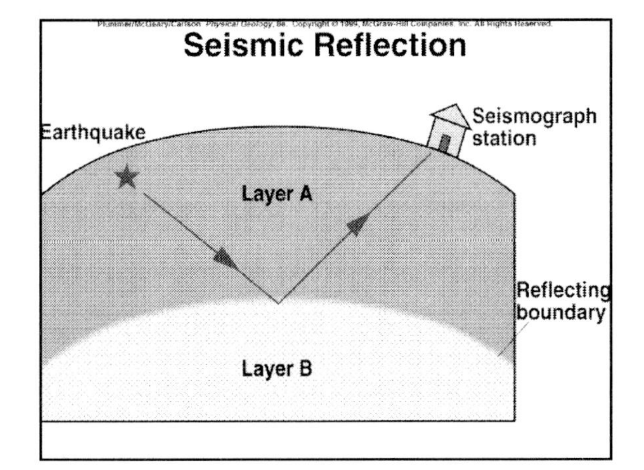

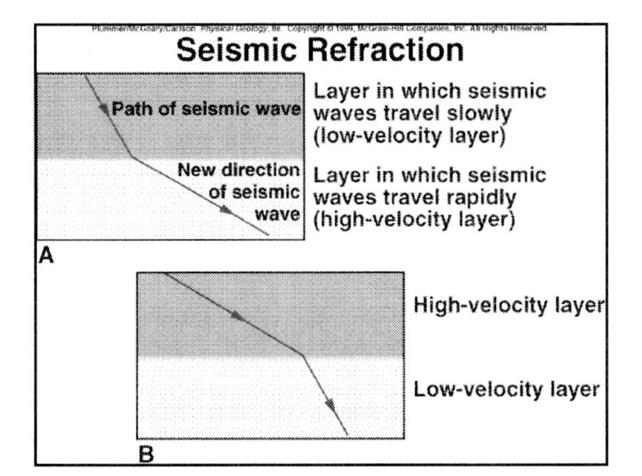

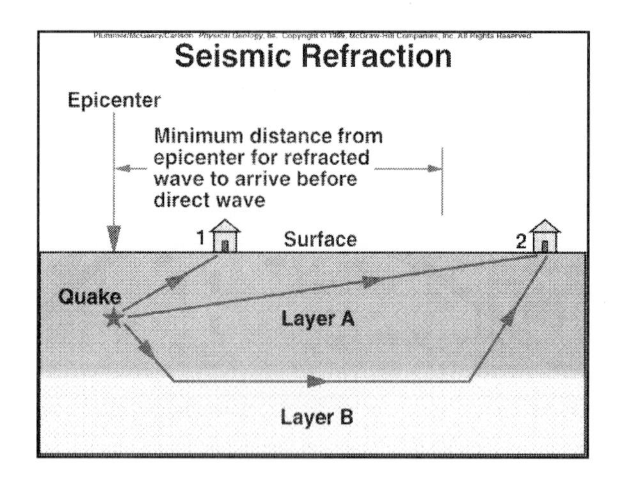
Plummer/McGeary/Carlson Physical Geology, 8e Copyright © 1999, McGraw-Hill Companies, Inc. All Rights Reserved.
Seismic Refraction
Epicenter
Minimum distance from epicenter for refracted wave to arrive before direct wave
1
Surface
2
Quake
Layer A
Layer B

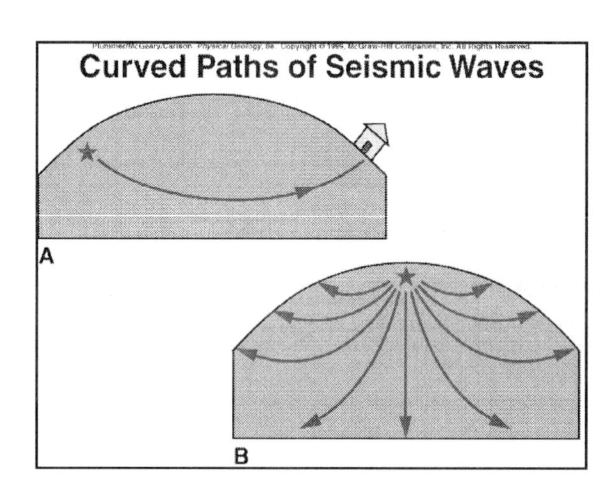
Plummer/McGeary/Carlson Physical Geology, 8e Copyright © 1999, McGraw-Hill Companies, Inc. All Rights Reserved.
Curved Paths of Seismic Waves
A
B

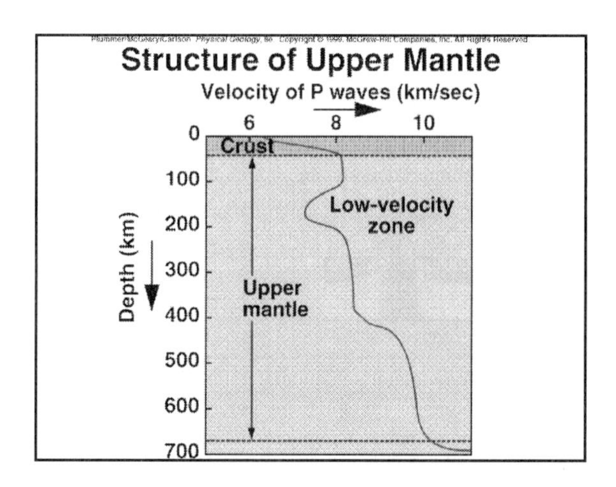
Plummer/McGeary/Carlson Physical Geology, 8e Copyright © 1999, McGraw-Hill Companies, Inc. All Rights Reserved.
Structure of Upper Mantle
Velocity of P waves (km/sec)
6
8
10
0
100
200
300
400
500
600
700
Depth (km)
Crust
Low-velocity zone
Upper mantle

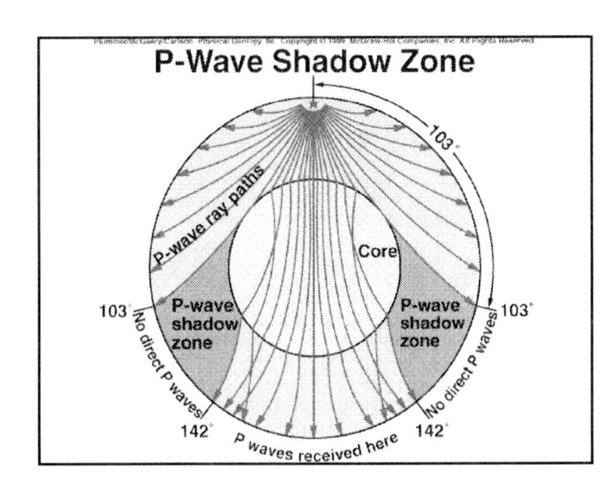
P-Wave Shadow Zone
103°
P-wave ray paths
Core
103°
P-wave shadow zone
P-wave shadow zone
103°
No direct P waves
No direct P waves
142°
P waves received here
142°

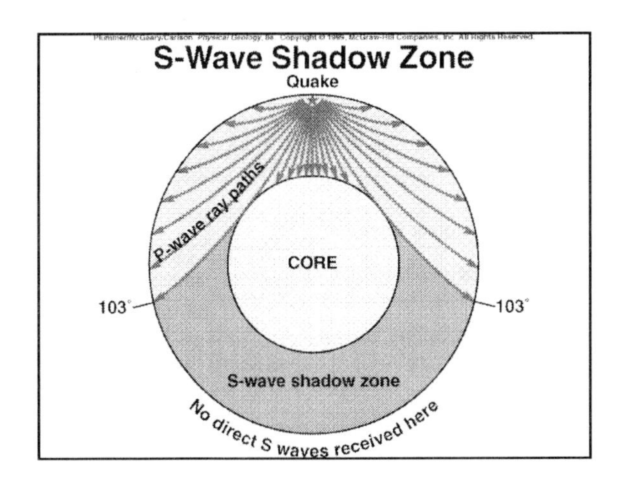
S-Wave Shadow Zone
Quake
P-wave ray paths
CORE
103°
103°
S-wave shadow zone
No direct S waves received here

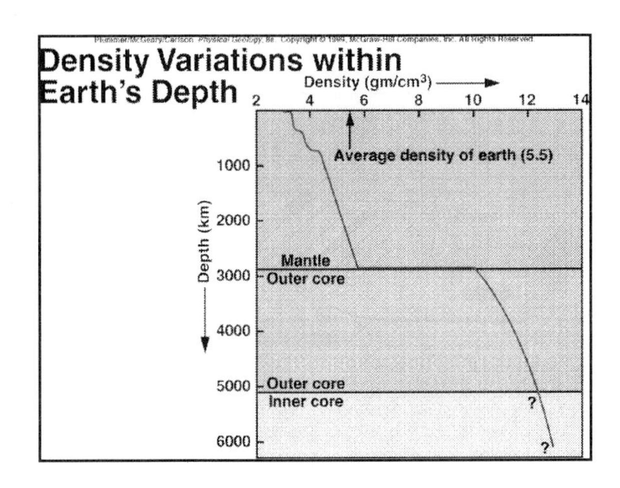
Density Variations within Earth's Depth
Density (gm/cm³)
2
4
6
8
10
12
14
Average density of earth (5.5)
1000
2000
3000
4000
5000
6000
Depth (km)
Mantle
Outer core
Outer core
Inner core
?
?

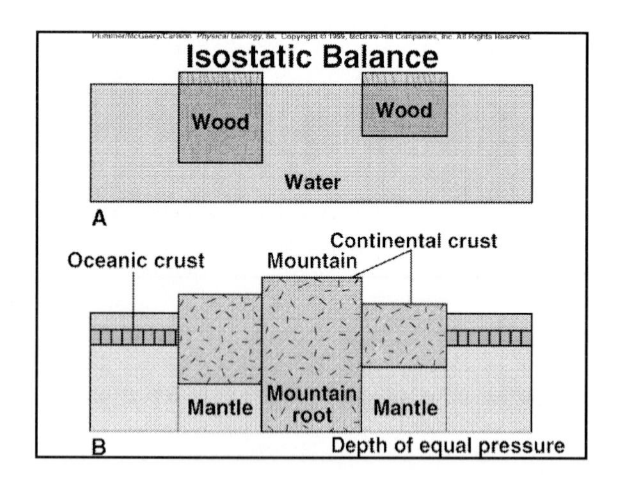

Isostatic Balance
Wood
Wood
Water
A
Continental crust
Oceanic crust
Mountain
Mantle
Mountain root
Mantle
B
Depth of equal pressure

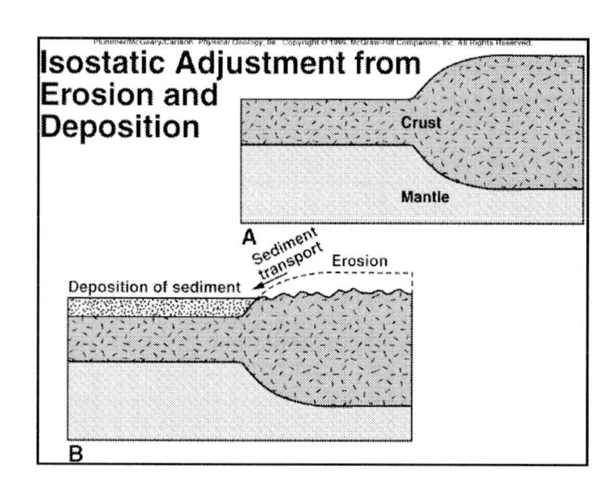

Isostatic Adjustment from Erosion and Deposition
Crust
Mantle
A
Sediment transport
Erosion
Deposition of sediment
B

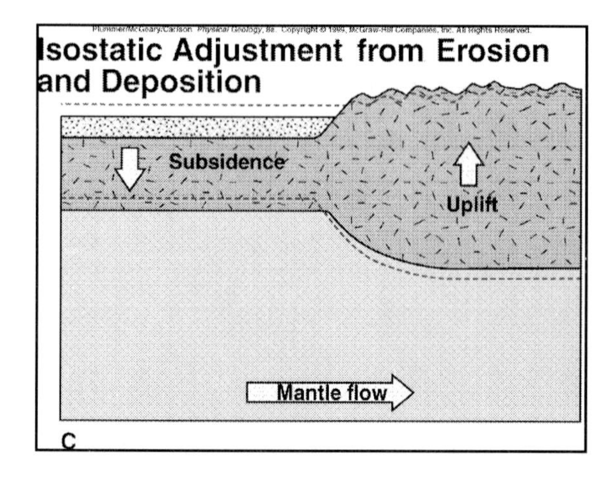

Isostatic Adjustment from Erosion and Deposition
Subsidence
Uplift
Mantle flow
C

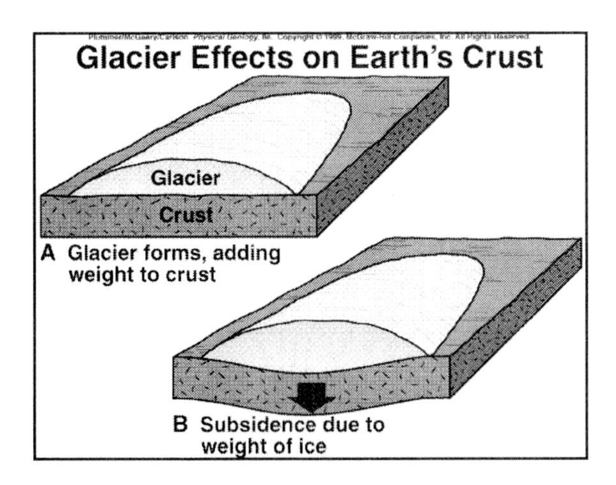
Glacier Effects on Earth's Crust
Glacier
Crust
A Glacier forms, adding weight to crust
B Subsidence due to weight of ice

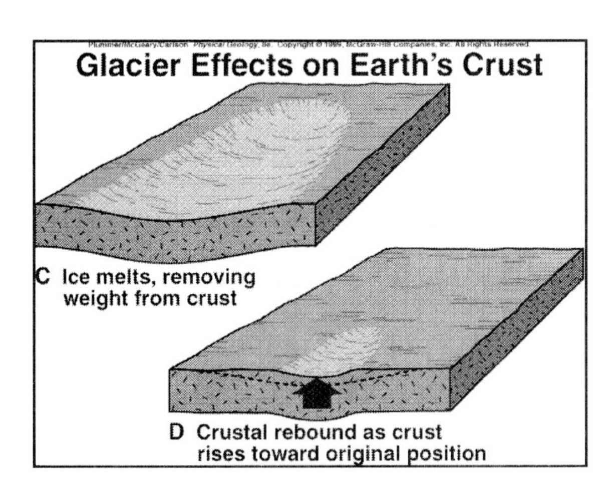
Glacier Effects on Earth's Crust
C Ice melts, removing weight from crust
D Crustal rebound as crust rises toward original position

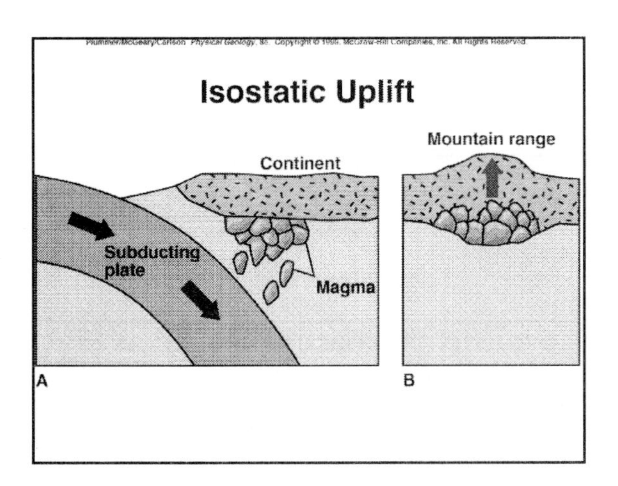
Isostatic Uplift
Continent
Mountain range
Subducting plate
Magma
A
B

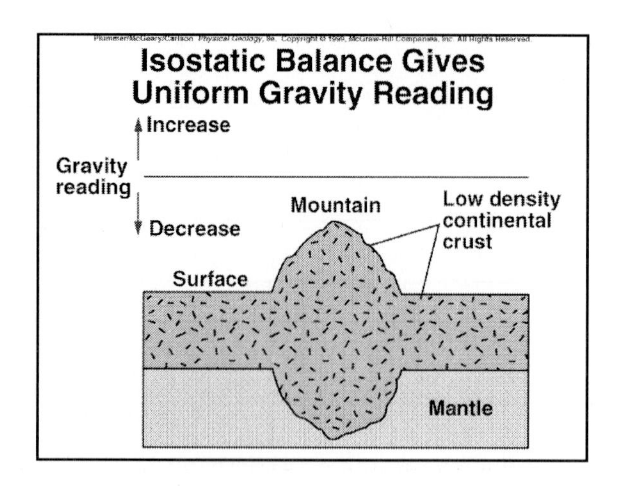
Isostatic Balance Gives Uniform Gravity Reading
Increase
Gravity reading
Decrease
Mountain
Low density continental crust
Surface
Mantle

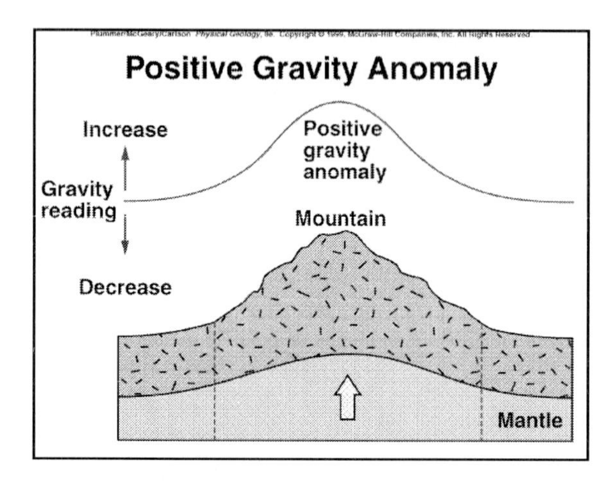
Positive Gravity Anomaly
Increase
Positive gravity anomaly
Gravity reading
Mountain
Decrease
Mantle

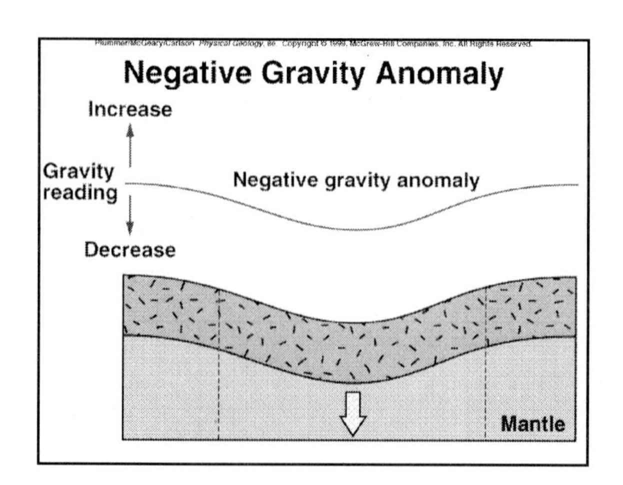
Negative Gravity Anomaly
Increase
Gravity reading
Negative gravity anomaly
Decrease
Mantle

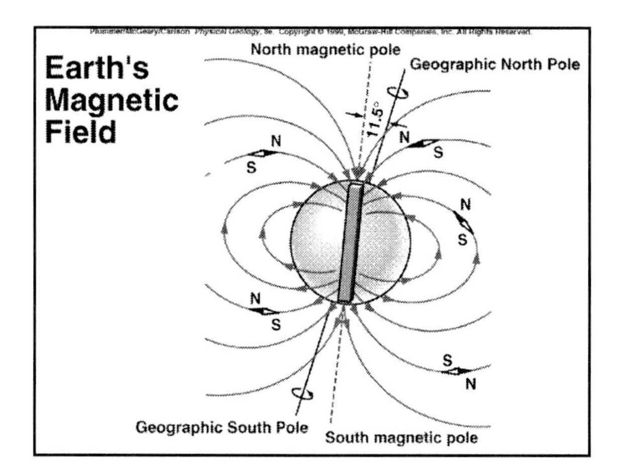
Plummer/McGeary/Carlson Physical Geology, 8e. Copyright © 1999, McGraw-Hill Companies, Inc. All Rights Reserved.
Earth's Magnetic Field
North magnetic pole
Geographic North Pole
11.5°
N
S
Geographic South Pole
South magnetic pole

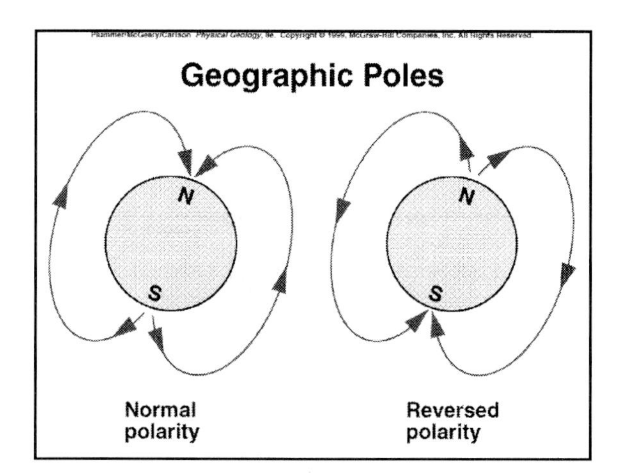
Plummer/McGeary/Carlson Physical Geology, 8e. Copyright © 1999, McGraw-Hill Companies, Inc. All Rights Reserved.
Geographic Poles
N
S
Normal polarity
Reversed polarity

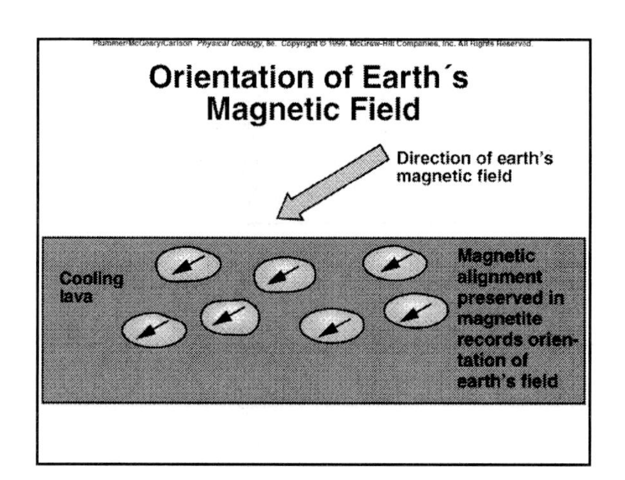
Plummer/McGeary/Carlson Physical Geology, 8e. Copyright © 1999, McGraw-Hill Companies, Inc. All Rights Reserved.
Orientation of Earth´s Magnetic Field
Direction of earth's magnetic field
Cooling lava
Magnetic alignment preserved in magnetite records orientation of earth's field

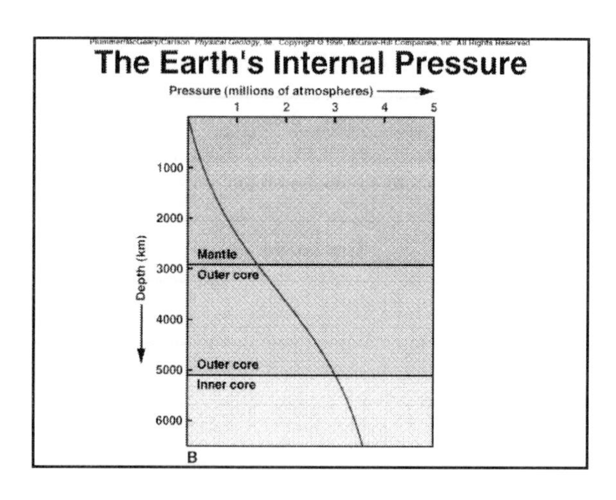
The Earth's Internal Pressure
Pressure (millions of atmospheres)
1
2
3
4
5
1000
2000
3000
4000
5000
6000
Depth (km)
Mantle
Outer core
Outer core
Inner core
B

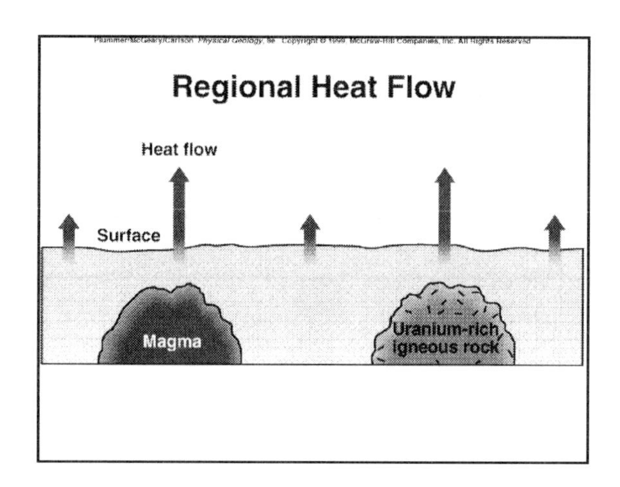
Regional Heat Flow
Heat flow
Surface
Magma
Uranium-rich igneous rock

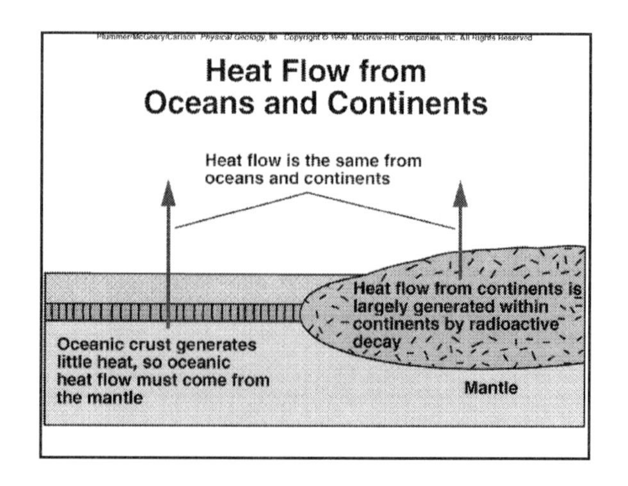
Heat Flow from Oceans and Continents
Heat flow is the same from oceans and continents
Heat flow from continents is largely generated within continents by radioactive decay
Oceanic crust generates little heat, so oceanic heat flow must come from the mantle
Mantle

Chapter 18

Physical Geology

eighth edition

Plummer/McGeary/Carlson

The Sea Floor

Origin of the Ocean

- Water vapor released during degassing of early earth
- Salt from chemical weathering

Methods of Studying the Sea Floor

- Rock Dredge
- Corer
- Sea-Floor Drilling
- Submersibles
- Echo Sounder
    - » Seismic Profiler
- Surveys - Magnetic, Gravity, Seismic Refraction
- Deep Sea Cameras

Features of the Sea Floor

- Continental Margins
    - Passive
    - Active
- Oceanic trench
- Mid-oceanic ridge
- Seamounts

Continental Shelves and Continental Slopes

- Vertical exaggeration in diagrams
- Continental shelf
- Continental slope

Submarine Canyons

- Abyssal Fans
- Bottom Currents
- Down-canyon movement of sand
- Bottom currents
- River erosion
- Turbidity Currents
  - Graded bedding
  - Shallow water fossils

Passive Continental Margins

- Continental shelf, slope, rise
- The Continental Rise
  - Types of Deposition
    - From turbidity currents
    - From contour currents
- Abyssal plains

Active Continental Margins

- On land- earthquakes, young mountain belt, volcanoes
- Continental shelf, continental slope, oceanic trench
- Oceanic Trenches
  - Earthquakes of the Benioff seismic Zones
  - Volcanoes
  - Low Heat Flow
  - Negative Gravity Anomalies

The Mid-Oceanic Ridge

- Rift Valley
- Geologic Activity on the Ridge
  - Shallow-focus Earthquakes
  - High Heat Flow
  - Basalt Eruptions
  - Hot springs
    - Black Smokers
- Biologic Activity on the Ridge
  - Geomicrobiology

Fracture Zones

- Offset rift valleys
- Transform Fault
  - Portion that has earthquakes

Seamounts, Guyots, and Aseismic Ridges

- Seamounts
- Guyots
- Aseismic ridges

Reefs

- Fringing Reefs
- Barrier Reefs
- Atolls

Sediments of the Sea Floor

- Terrigenous Sediment
- Pelagic Sediment
  - thickness increases away from crest of mid-oceanic ridge

Oceanic Crust and Ophiolites

- Evidence for compostion of the oceanic crust
- Ophiolite (from top to bottom)
  - Marine sedimentary rock
  - Pillow basalt
  - Sheeted dike complex
  - Gabbroic intrusions
  - Ultramafic rock

The Age of the Sea Floor

- Younger than 200 million years old
- Parts of continents much older

The Sea Floor and Plate Tectonics

- Material in this chapter is *data*
- Next chapter provides hypotheses and theories to explain the data
- Most can be explained through plate tectonics

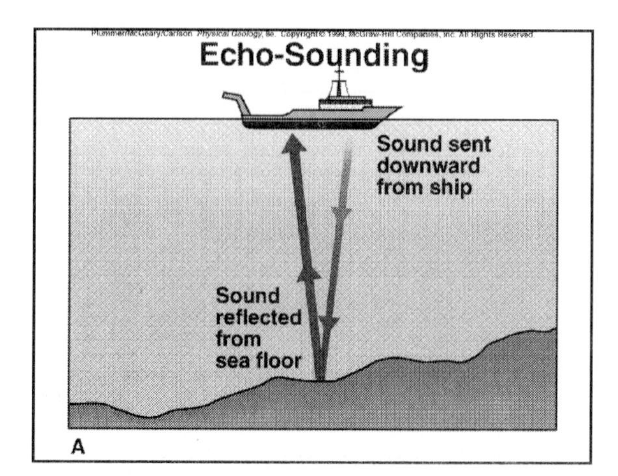

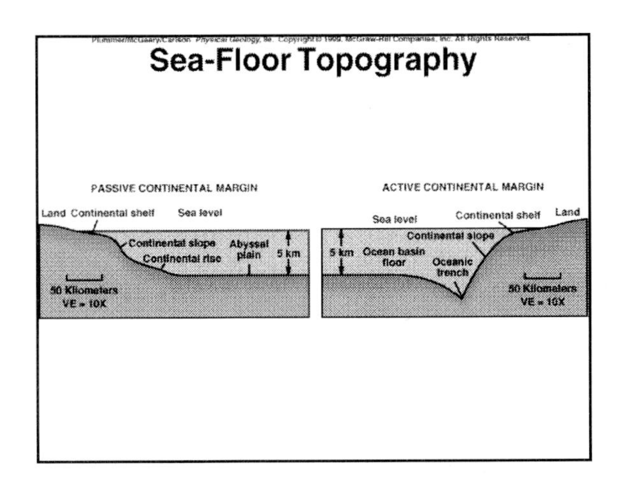
Sea-Floor Topography
PASSIVE CONTINENTAL MARGIN
Land
Continental shelf
Sea level
Continental slope
Continental rise
Abyssal plain
5 km
50 Kilometers
VE = 10X
ACTIVE CONTINENTAL MARGIN
Sea level
Continental shelf
Land
Continental slope
5 km
Ocean basin floor
Oceanic trench
50 Kilometers
VE = 10X

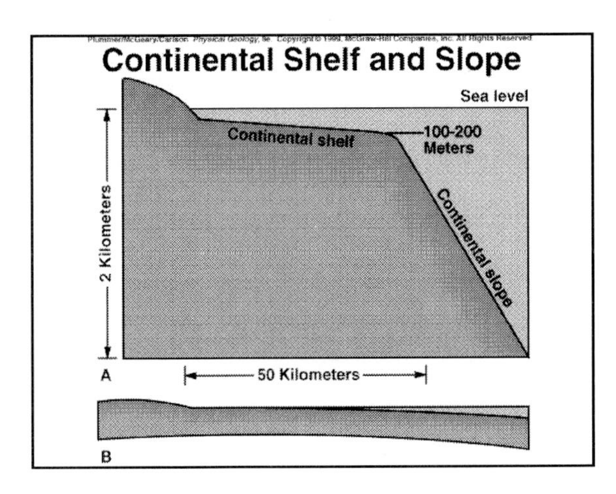
Continental Shelf and Slope
Sea level
Continental shelf
100-200 Meters
Continental slope
2 Kilometers
A
50 Kilometers
B

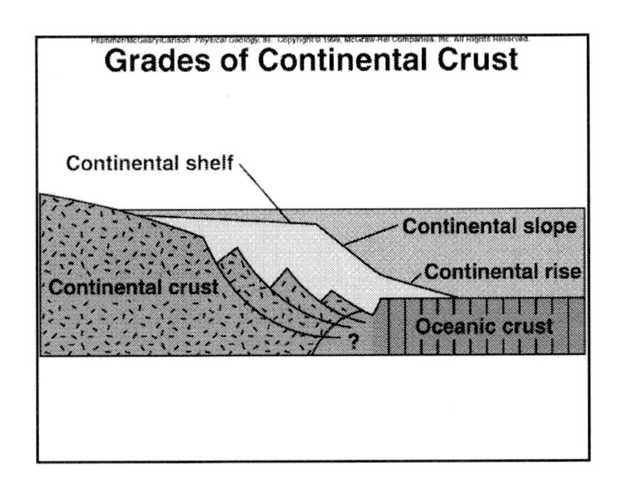
Grades of Continental Crust
Continental shelf
Continental slope
Continental rise
Continental crust
Oceanic crust
?

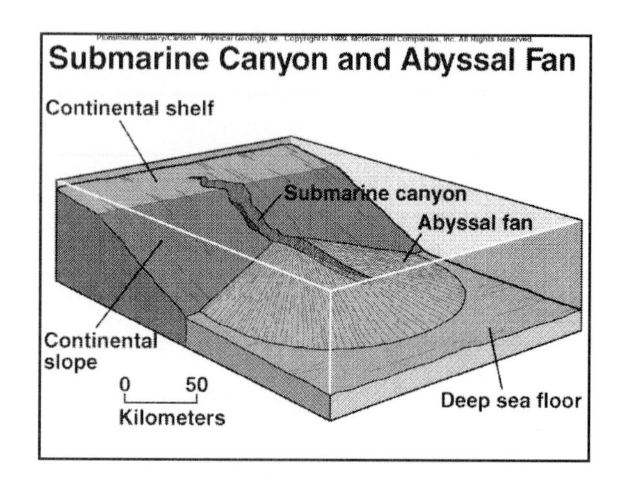
Plummer/McGeary/Carlson Physical Geology, 8e Copyright © 1999, McGraw-Hill Companies, Inc. All Rights Reserved
Submarine Canyon and Abyssal Fan
Continental shelf
Submarine canyon
Abyssal fan
Continental slope
0 50
Kilometers
Deep sea floor

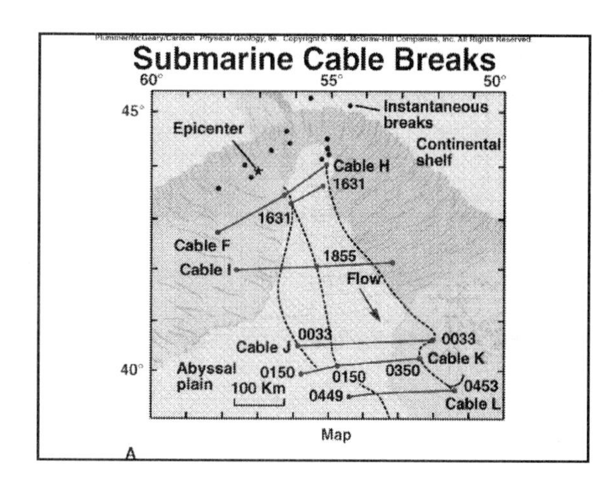
Plummer/McGeary/Carlson Physical Geology, 8e Copyright © 1999, McGraw-Hill Companies, Inc. All Rights Reserved
Submarine Cable Breaks
60°
55°
50°
45°
40°
Instantaneous breaks
Epicenter
Continental shelf
Cable H
1631
1631
Cable F
Cable I
1855
Flow
0033
0033
Cable J
Cable K
Abyssal plain
0150
0150
0350
0453
100 Km
0449
Cable L
Map
A

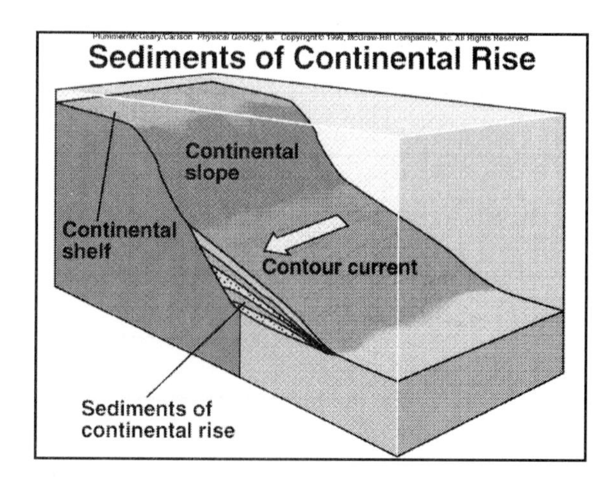
Plummer/McGeary/Carlson Physical Geology, 8e Copyright © 1999, McGraw-Hill Companies, Inc. All Rights Reserved
Sediments of Continental Rise
Continental slope
Continental shelf
Contour current
Sediments of continental rise

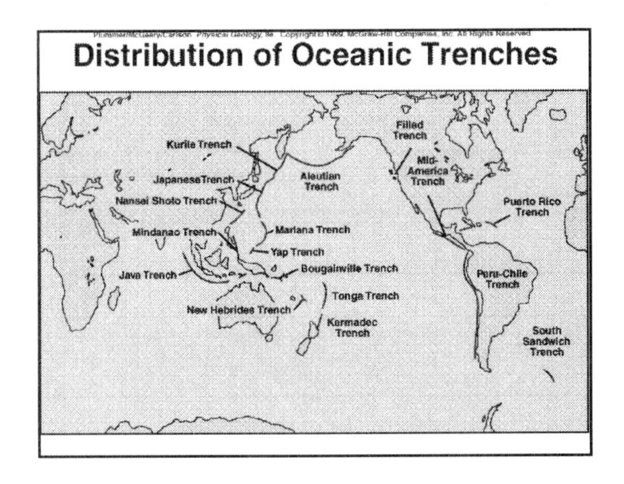
Distribution of Oceanic Trenches
Kurile Trench
Japanese Trench
Nansei Shoto Trench
Mindanao Trench
Java Trench
New Hebrides Trench
Aleutian Trench
Mariana Trench
Yap Trench
Bougainville Trench
Tonga Trench
Kermadec Trench
Filled Trench
Mid-America Trench
Puerto Rico Trench
Peru-Chile Trench
South Sandwich Trench

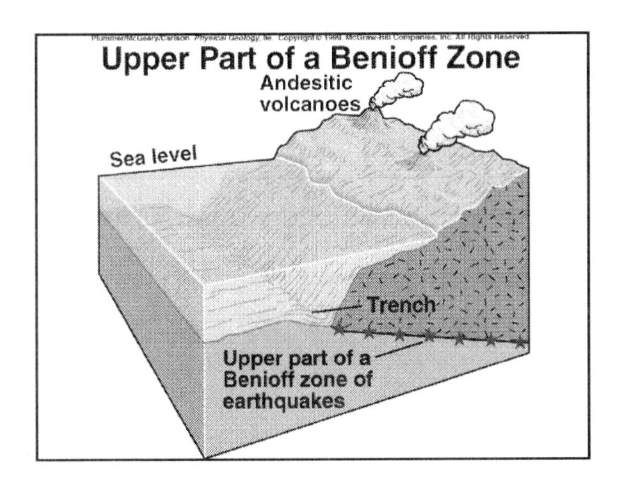
Upper Part of a Benioff Zone
Andesitic volcanoes
Sea level
Trench
Upper part of a Benioff zone of earthquakes

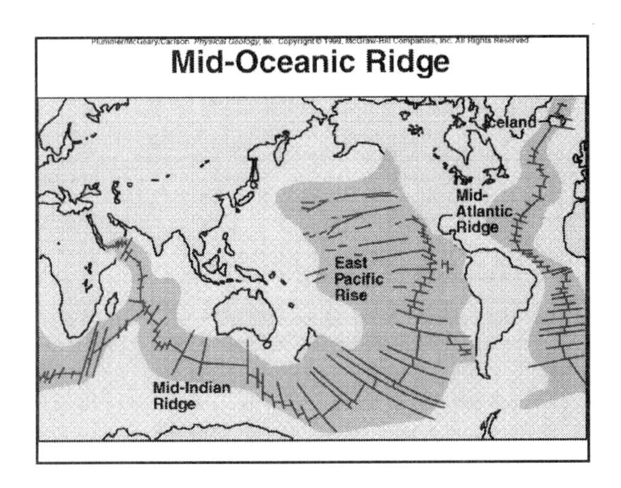
Mid-Oceanic Ridge
Iceland
Mid-Atlantic Ridge
East Pacific Rise
Mid-Indian Ridge

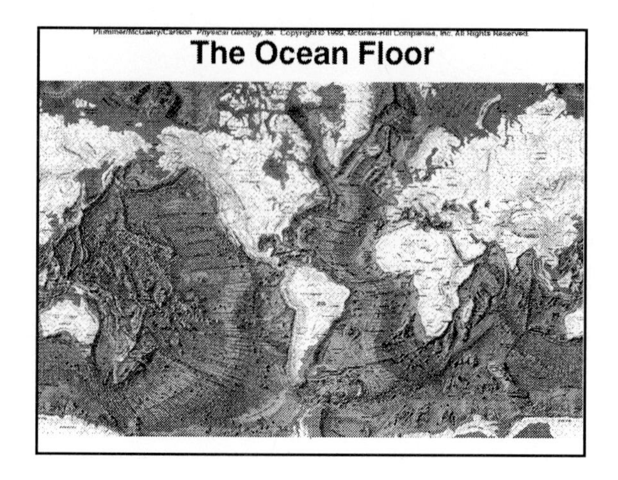
The Ocean Floor

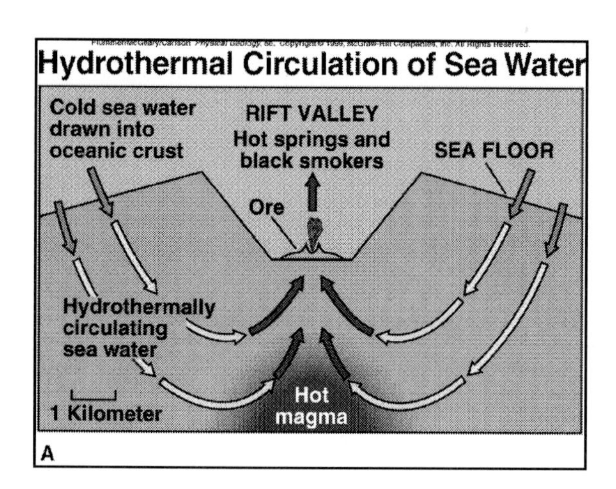
Hydrothermal Circulation of Sea Water
Cold sea water drawn into oceanic crust
RIFT VALLEY
Hot springs and black smokers
SEA FLOOR
Ore
Hydrothermally circulating sea water
1 Kilometer
Hot magma
A

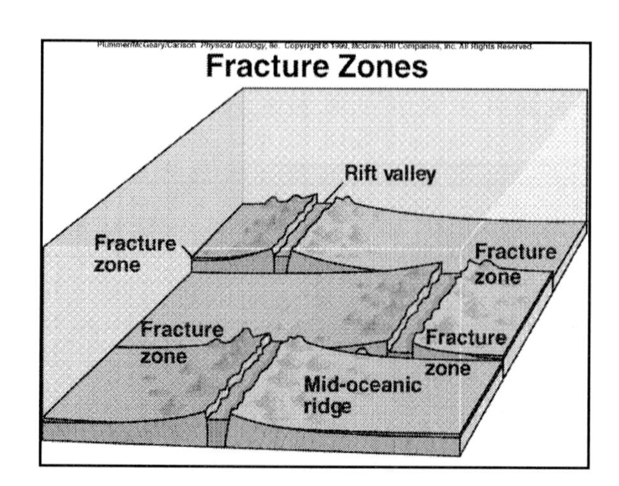
Fracture Zones
Rift valley
Fracture zone
Fracture zone
Fracture zone
Fracture zone
Mid-oceanic ridge

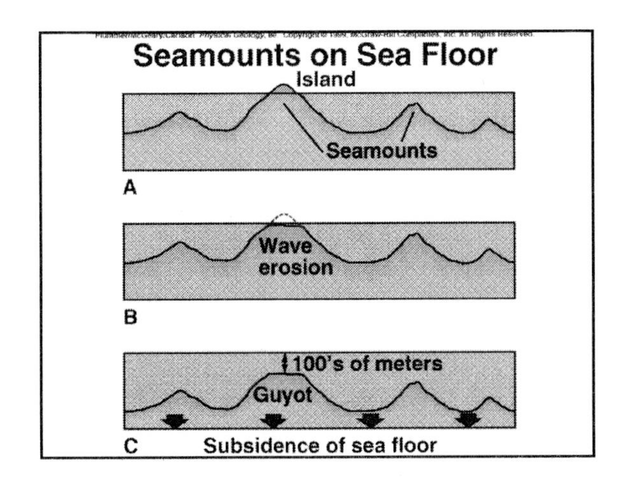
Seamounts on Sea Floor
Island
Seamounts
A
Wave
erosion
B
100's of meters
Guyot
C
Subsidence of sea floor

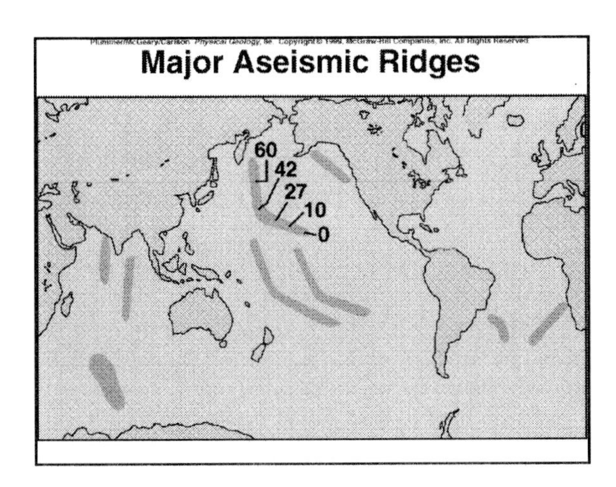
Major Aseismic Ridges
60
42
27
10
0

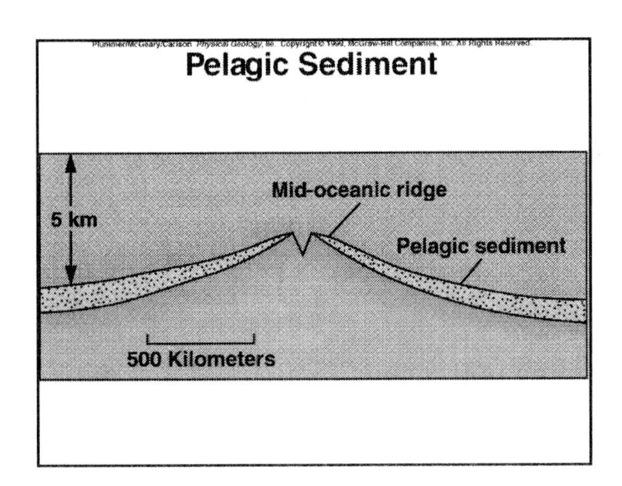
Pelagic Sediment
5 km
Mid-oceanic ridge
Pelagic sediment
500 Kilometers

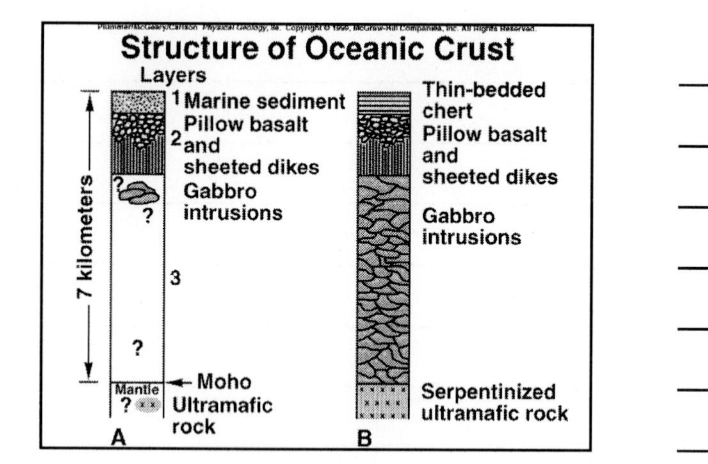
Structure of Oceanic Crust
Layers
7 kilometers
1 Marine sediment
2 Pillow basalt and sheeted dikes
Gabbro intrusions
3
?
Mantle
Moho
Ultramafic rock
A
Thin-bedded chert
Pillow basalt and sheeted dikes
Gabbro intrusions
Serpentinized ultramafic rock
B

Chapter 19

Physical Geology

eighth edition

Plummer/McGeary/Carlson

PLATE TECTONICS

INTRODUCTION

- *Tectonics*
- Plate tectonics
  - Earth's outer shell divided into plates
  - Plates move & change in size
- Activity at plate boundaries
- Combined:
  - Continental drift
  - Sea-floor spreading

Early Case for Continental Drift

- Continental coastlines would fit together
- Rocks & fossils indicated that continents joined
  - *Pangea*- supercontinent
    - Separated into *Laurasia* & *Gondwanaland*
- Late Paleozoic glaciation
- Paleoclimatology indicated *polar wandering*
- Skepticism about Continental Drift
  - Problem of forces

Paleomagnetism & Revival of Continental Drift

- Magnetite aligns on existing magnetic field
- Dip indicates old magnetic pole position
- Apparent motion of north magnetic pole through time
  - Split in path
  - indicates continents split apart

Recent Evidence for Continental Drift

- Fitting continents at continental slope rather than shoreline
- Refined matches of rocks between continents
- Isotopic ages support matches
- Glacial evidence
- Matches between Africa and South America are particularly convincing

History of Continental Positions

- Pangea split up 200 m.y.
- Continents in motion for at least 2 billion years

SEA-FLOOR SPREADING

- Sea-floor moves away from mid-oceanic ridge
- Plunges beneath continent or island arc- *subduction*
- Rate of 1 to 6 (or more) cm/year
- Driving force
  - Originally regarded as mantle convection

SEA-FLOOR SPREADING

- Explanations
  - Mid-oceanic ridge
    - Hot mantle rock beneath ridge
      - High heat flow
      - Basalt eruptions
    - Rift valley
    - Shallow-focus earthquakes

SEA-FLOOR SPREADING

- Explanations
  - Oceanic trenches
    - » Low heat flow
    - » Negative gravity anomalies
    - » Benioff zone earthquakes
    - » Andesitic volcanism
  - Age of sea floor
    - » Young age of sea floor rocks
    - » Implies youngest should be at ridges, oldest at trenches
    - » Explains pattern of pelagic sediment

Plates and Plate Motion

- *Plate*
  - Entirely sea floor or
  - continental and oceanic
- *Lithosphere*
  - Crust & uppermost mantle
  - Thickness increases away from ridge
- *Asthenosphere*
  - Low seismic velocity zone
  - behaves plastically

Plates and Plate Motion

- Interior of plates relatively inactive
- Activity along boundaries
  - e.g., earthquakes, volcanoes, young mountain belts
- Plate tectonics a unifying theory for geology
- Boundaries
  - Divergent, Convergent, Transform

How do we know that plates move?

- Marine magnetic anomalies
  - Vine-Matthews Hypothesis
    - » Anomalies
    - » Reversals
    - » Normal and reverse polarity
    - » Positive and negative anomalies
  - Measuring the rate of sea floor
  - Predicting sea floor age

How do we know that plates move?

- Fracture Zones & Transform Faults
  - Pattern of earthquakes at ridges and fracture zones
  - Transform fault
- Measuring plate motion directly
  - Use of satellites

DIVERGENT BOUNDARIES

- During break up of a continent
  - Rifting, basaltic eruptions, uplift
  - Extension- normal faults, rift valley (graben) forms
  - Shallow focus earthquakes
- Continental crust separates
  - Fault blocks along edges
  - Oceanic crust created
  - Rock salt may develop in rift

DIVERGENT BOUNDARIES

- Continuing divergence
  - Widening sea
  - Mid-oceanic ridge
  - Marine sediment covers continental edges
  - Passive continental margin
  - New crust formed at mid-oceanic ridge
    - Pillow basalt and dikes

TRANSFORM BOUNDARIES

- Two plates slide past each other
- Usually between mid-oceanic ridge segments
  - Can also connect ridge and trench
  - Or trench to trench
- Origin of offset of ridges

CONVERGENT BOUNDARIES

- Plates move toward each other
- One plate overrides the other
  - *Subduction zone*

CONVERGENT BOUNDARIES

- Oceanic-Oceanic Convergence
  - Oceanic trench
    - » curved convex to subducting plate
  - Beniofff zone
  - Magma generated at depth
    - » Andesitic volcanism
  - Island arc forms
    - » Angle of subduction determines distance of arc from trench
  - Accretionary wedge
  - trench migration in time

CONVERGENT BOUNDARIES

- Oceanic-Continental Convergence
  - Active continental margin
    - » Subduction of oceanic lithosphere beneath continental lithosphere
    - » Accretionary wedge & forearc basin
  - Magmatic arc- volcanoes & plutons
  - Crustal thickening and mountain belts
  - Regional metamorphism
  - Thrust faulting & folding on continental side
    - » Backarc basin

CONVERGENT BOUNDARIES

- Continental-Continental convergence
  - Two continents approach each other and collide
    - » Sea floor subducted on one side
    - » Ocean becomes narrower and narrower
    - » Continent wedged into subduction zone but not carried down it
    - » Suture zone
  - Crust thickened
    - » Two thrust belts
  - Mountain belt in interior of continent

Backarc spreading

- Regional extension behind arc
- New oceanic crust created
- Possible causes

Motion of Plate boundaries

- Boundaries move as well as plates
- Ridge crests may jump to a new position
- Convergent boundaries can migrate or jump
- Transform boundaries can change position
  - San Andreas fault may shift

Plate size

- New sea floor added to trailing edge of plate
  - e.g. North American plate growing at mid-Atlantic ridge
- Oceanic plate might get smaller as continetal plate overrides it
  - e.g. Nazca plate subducted beneath westward moving South American plate

Attractiveness of Plate Tectonics

- Many of earth features explained. Summary:
  - Distribution of volcanoes
    - Basaltic at diverging boundaries
    - Andesitic at converging boundaries
  - Earthquake distribution
  - Young mountain belts
  - Sea floor
    - Mid-oceanic ridge
    - Oceanic trenches
    - Fracture zones

What Causes Plate Motions?

- Need to explain:
  - Mid-oceanic ridges hot & elevated; trenches cold & deep
  - Ridge crests have tensional cracks
  - Leading edges of some plates subducting sea floor
  - Leading edges of other plates are continents
- Convection in mantle (3 hypotheses)
  - Deep mantle convection
  - Two-layer convection

What Causes Plate Motions?

- Convection in mantle
  - Convection a *result* of plate motion
    - Ridge push
    - Slab pull
    - Trench suction

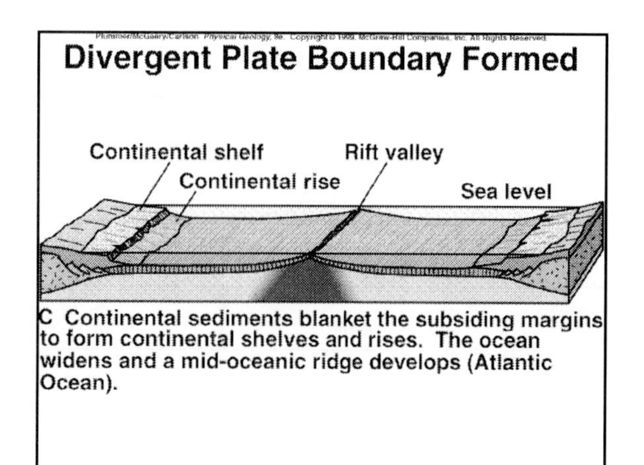
Divergent Plate Boundary Formed
Continental shelf
Continental rise
Rift valley
Sea level
C Continental sediments blanket the subsiding margins to form continental shelves and rises. The ocean widens and a mid-oceanic ridge develops (Atlantic Ocean).

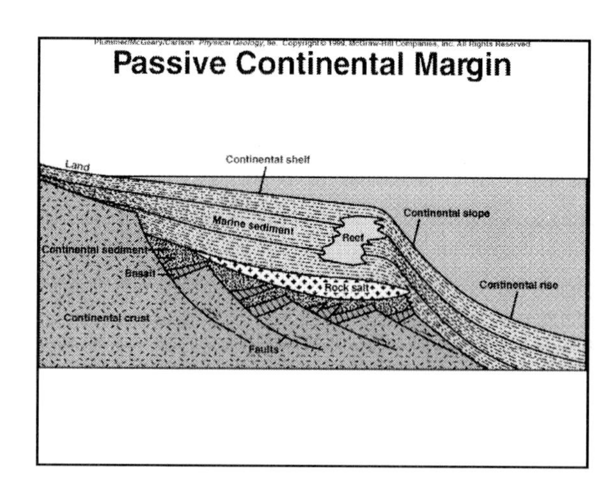
Passive Continental Margin
Land
Continental shelf
Marine sediment
Reef
Continental slope
Continental sediment
Basalt
Rock salt
Continental rise
Continental crust
Faults

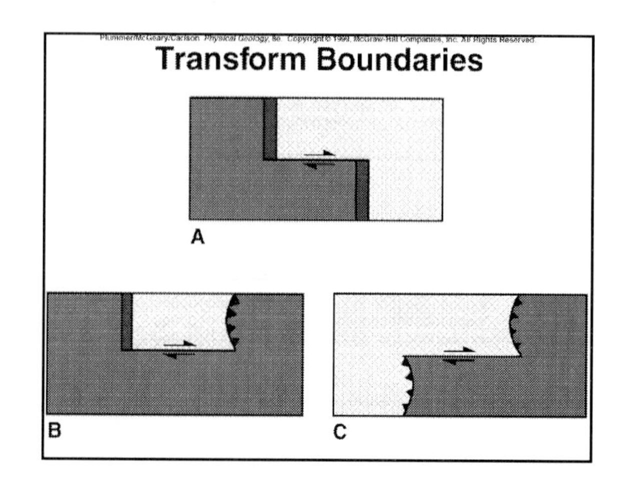
Transform Boundaries
A
B
C

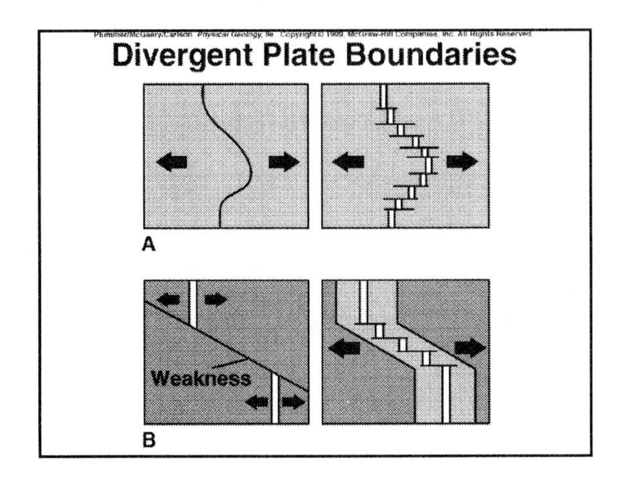
Plummer/McGeary/Carlson Physical Geology, 8e Copyright © 1999 McGraw-Hill Companies, Inc. All Rights Reserved
Divergent Plate Boundaries
A
Weakness
B

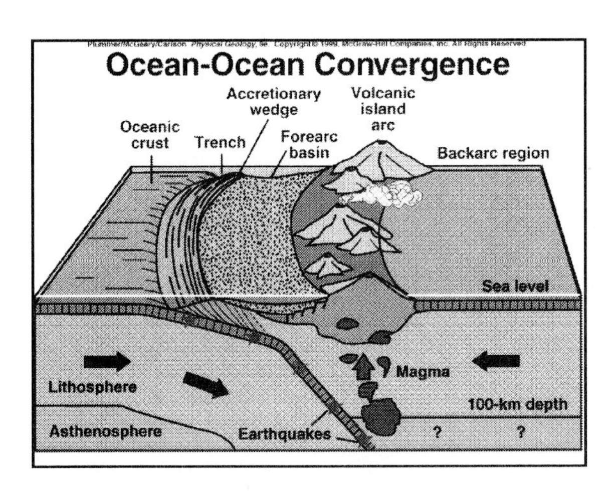
Ocean-Ocean Convergence
Accretionary wedge
Volcanic island arc
Oceanic crust
Trench
Forearc basin
Backarc region
Sea level
Lithosphere
Magma
100-km depth
Asthenosphere
Earthquakes
?
?

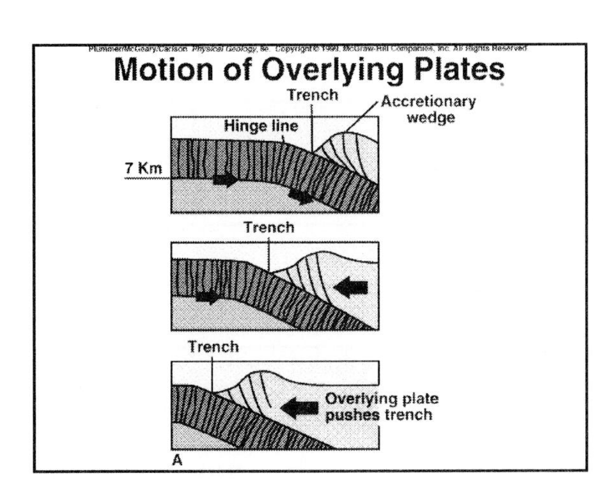
Motion of Overlying Plates
Trench
Accretionary wedge
Hinge line
7 Km
Trench
Trench
Overlying plate pushes trench
A

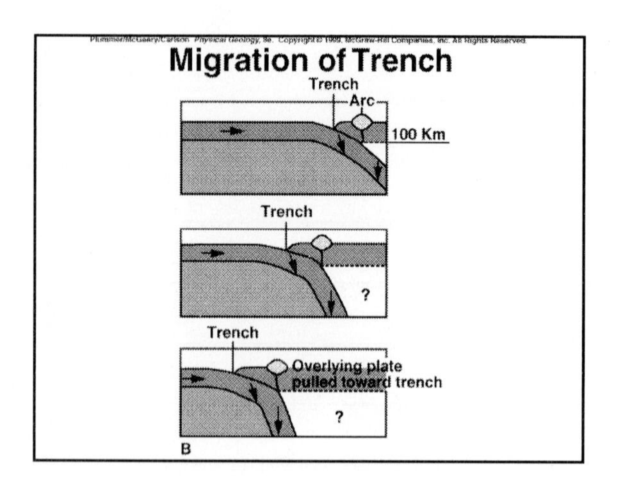
Migration of Trench
Trench
Arc
100 Km
Trench
?
Trench
Overlying plate pulled toward trench
?
B

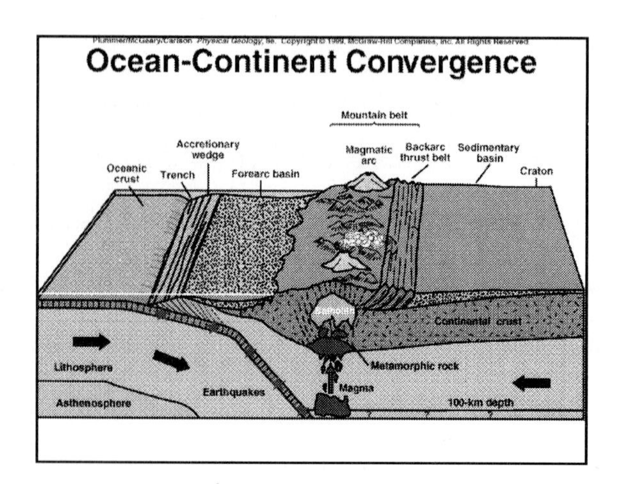
Ocean-Continent Convergence
Mountain belt
Accretionary wedge
Magmatic arc
Backarc thrust belt
Sedimentary basin
Oceanic crust
Trench
Forearc basin
Craton
Batholith
Continental crust
Lithosphere
Metamorphic rock
Magma
Earthquakes
Asthenosphere
100-km depth

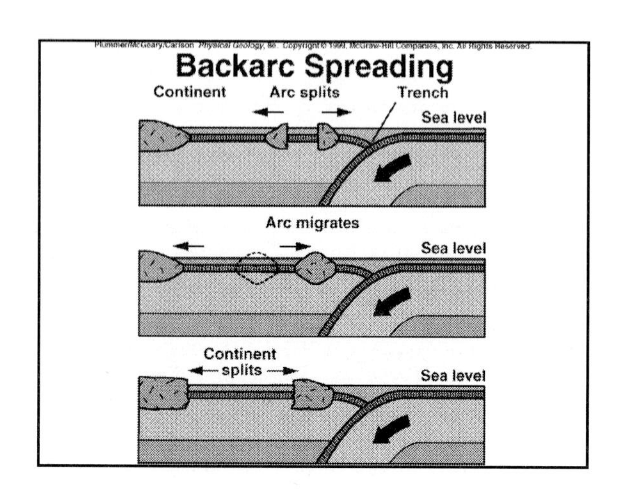
Backarc Spreading
Continent
Arc splits
Trench
Sea level
Arc migrates
Sea level
Continent splits
Sea level

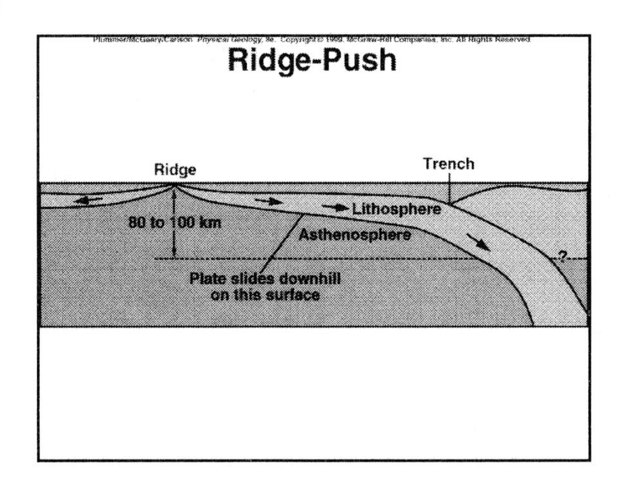
Ridge-Push
Ridge
Trench
Lithosphere
80 to 100 km
Asthenosphere
?
Plate slides downhill
on this surface

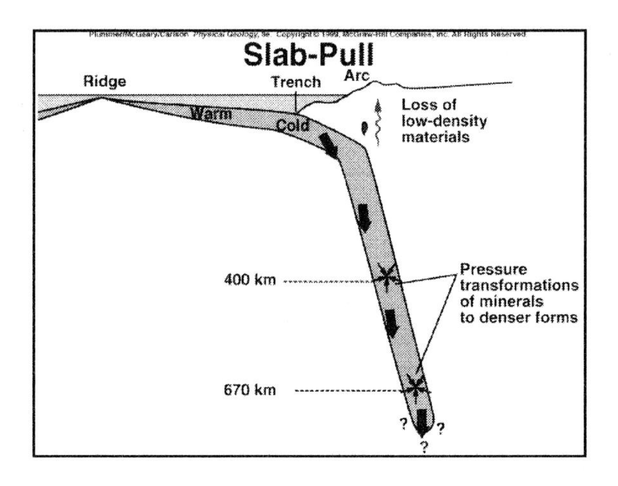
Slab-Pull
Ridge
Trench
Arc
Warm
Cold
Loss of
low-density
materials
400 km
Pressure
transformations
of minerals
to denser forms
670 km
?
?
?

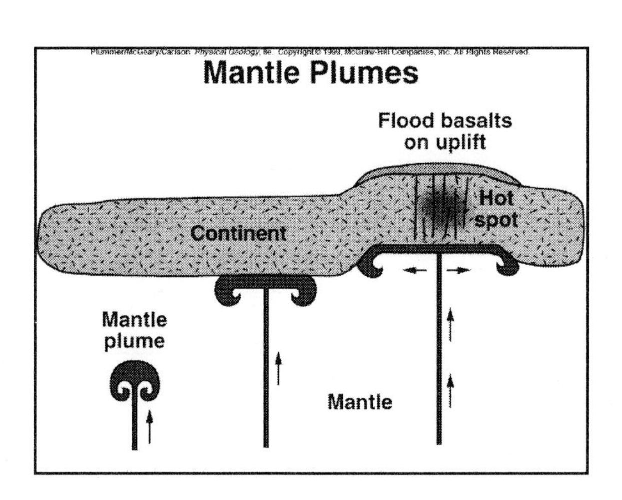
Mantle Plumes
Flood basalts
on uplift
Hot
spot
Continent
Mantle
plume
Mantle

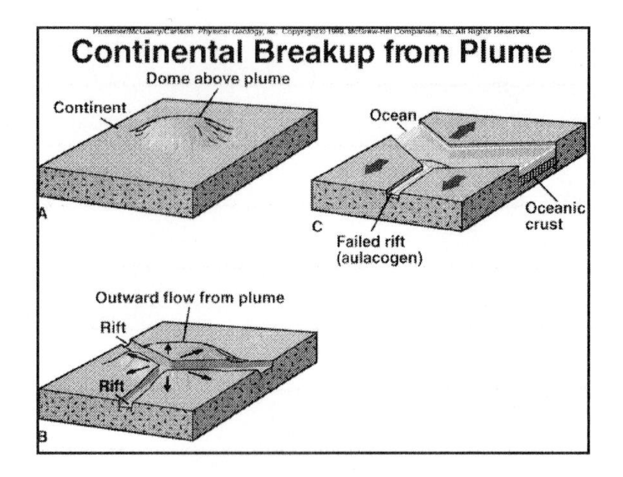
Continental Breakup from Plume
Dome above plume
Continent
A
Ocean
Oceanic crust
C
Failed rift (aulacogen)
Outward flow from plume
Rift
Rift
B

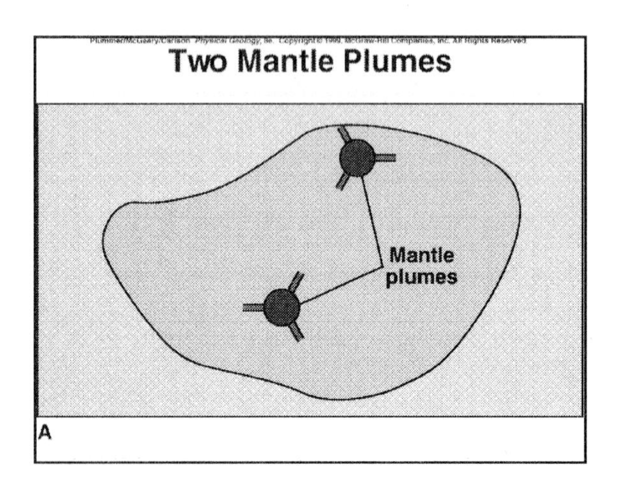
Two Mantle Plumes
Mantle plumes
A

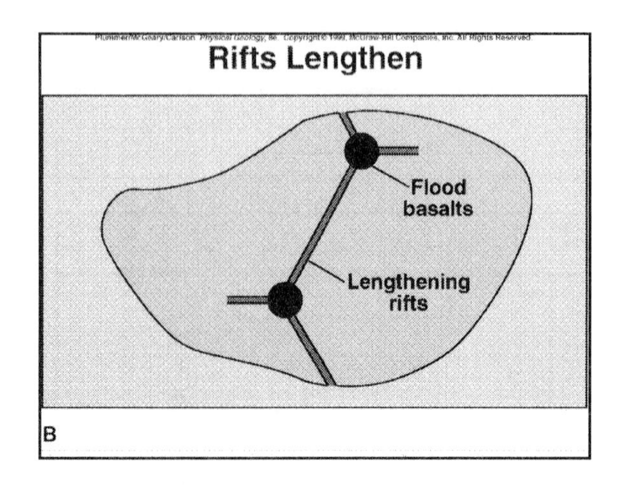
Rifts Lengthen
Flood basalts
Lengthening rifts
B

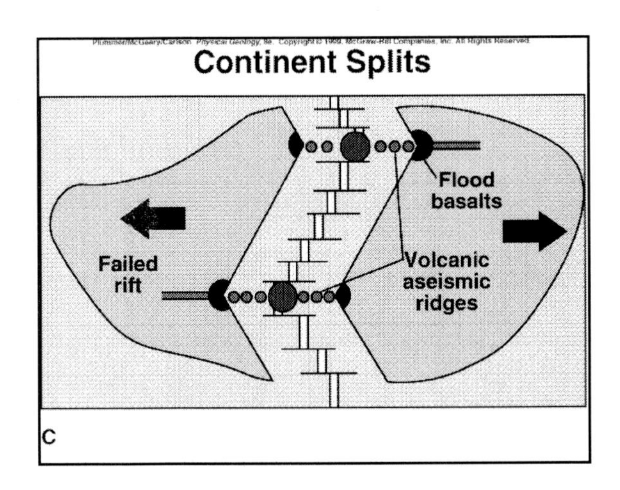
Continent Splits
Flood basalts
Failed rift
Volcanic aseismic ridges
C

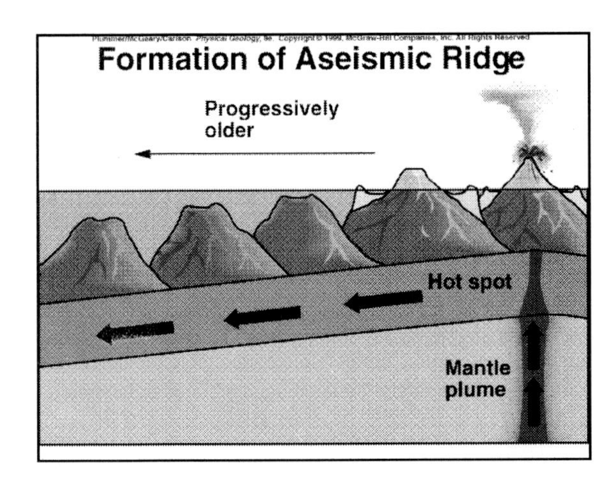
Formation of Aseismic Ridge
Progressively older
Hot spot
Mantle plume

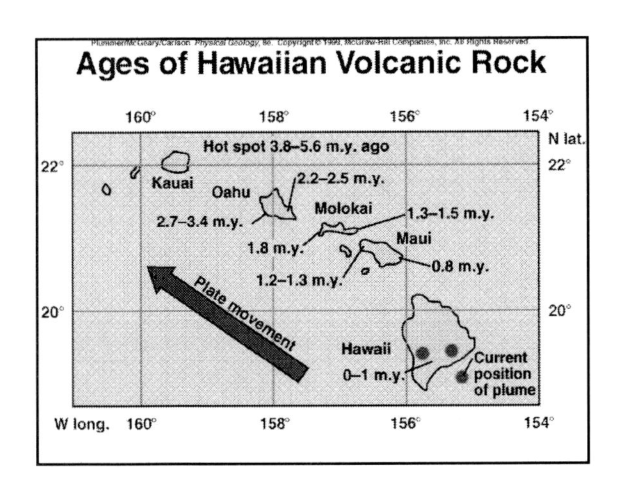
Ages of Hawaiian Volcanic Rock
160°
158°
156°
154°
N lat.
22°
20°
Hot spot 3.8–5.6 m.y. ago
Kauai
Oahu
2.2–2.5 m.y.
2.7–3.4 m.y.
Molokai
1.3–1.5 m.y.
1.8 m.y.
Maui
0.8 m.y.
1.2–1.3 m.y.
Plate movement
Hawaii
0–1 m.y.
Current position of plume
W long.

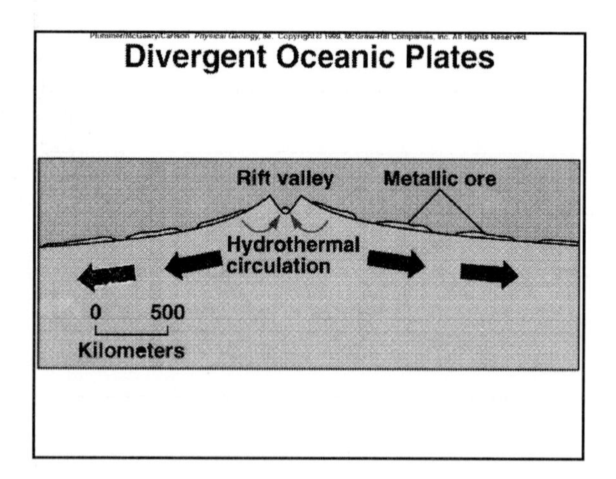
Plummer/McGeary/Carlson Physical Geology, 8e. Copyright © 1999, McGraw-Hill Companies, Inc. All Rights Reserved.
Divergent Oceanic Plates
Rift valley
Metallic ore
Hydrothermal
circulation
0
500
Kilometers

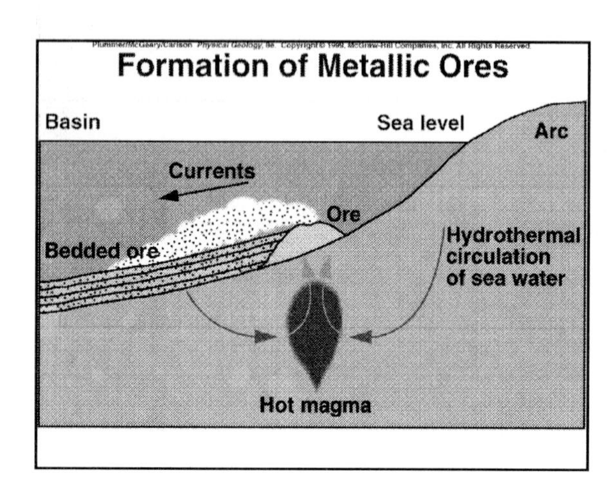
Plummer/McGeary/Carlson Physical Geology, 8e. Copyright © 1999, McGraw-Hill Companies, Inc. All Rights Reserved.
Formation of Metallic Ores
Basin
Sea level
Arc
Currents
Ore
Bedded ore
Hydrothermal
circulation
of sea water
Hot magma

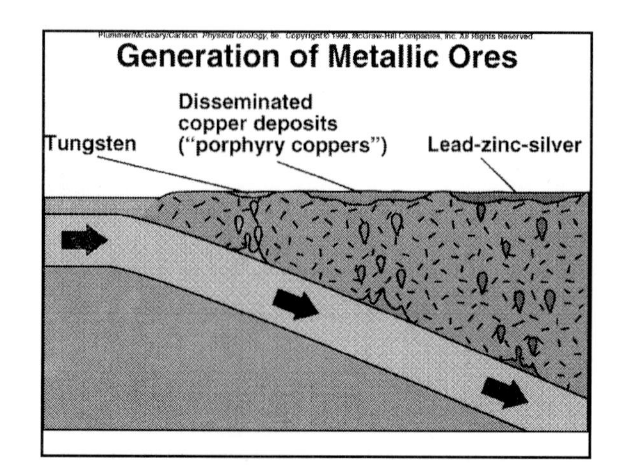
Plummer/McGeary/Carlson Physical Geology, 8e. Copyright © 1999, McGraw-Hill Companies, Inc. All Rights Reserved.
Generation of Metallic Ores
Disseminated
copper deposits
("porphyry coppers")
Tungsten
Lead-zinc-silver

Chapter 20

Physical Geology

eighth edition

Plummer/McGeary/Carlson

Mountain Belts and the Continental Crust

Introduction

- Mountain
- Major mountain belts
- Mountain range

Characteristics of Major Mountain Belts

- Size and Alignment
- Ages of Mountain Belts and Continents
  - Youngest tend to be higher
  - Craton
    - Precambrian Shield
- Thickness and Characteristics of Rock Layers

Characteristics of Major Mountain Belts

- Patterns of Flooding and Faulting
  - Fold and Thrust Belts
    - Detachment Fault
    - Crustal Shortening
    - Crustal Thickening
- Metamorphism and Plutonism
- Normal Faulting
- Thickness and Density of Rocks
- Features of Active Mountain Ranges

Evolution of a Mountain Belt

- Accumulation Stage
  - Accumulation in an Opening Ocean Basin
  - Accumulation along a Convergent Boundary
    - Graywackes
    - Magmatic Arc

Evolution of a Mountain Belt

- The Orogenic Stage
  - Orogeny
  - Orogenies and Ocean-Continent Convergence
    - Accretionary Wedge
    - Gravitational Collapse and Spreading
  - Arc-Continent Convergence
  - Orogenies and Continent-Continent Convergence
    - Wilson Cycle

Evolution of a Mountain Belt

- The Uplift and Block-faulting Stage
  - Isostacy
    - Isostatic Adjustment
  - Normal Faulting
    - Fault-block Mountain Range
  - Delamination

The Growth of Continents

- Crust added by accumulation & igneous activity
- Suspect and Exotic Terranes
  - Tectonostratigraphic Terranes- "Terranes"
  - Suspect terrane
  - Accreted terrane
    - Exotic terrane

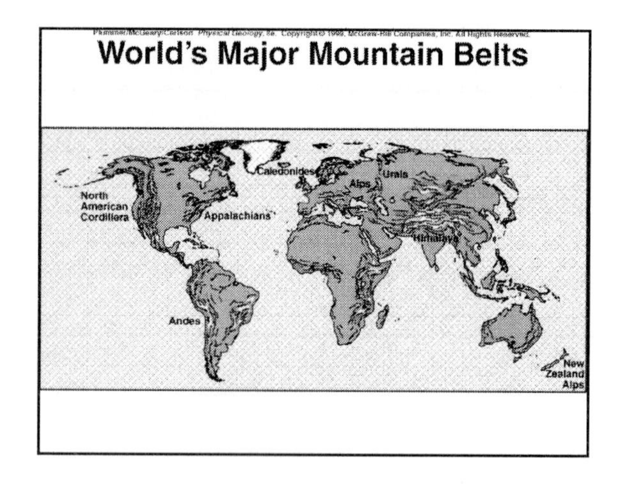
World's Major Mountain Belts
North American Cordillera
Appalachians
Caledonides
Alps
Urals
Himalaya
Andes
New Zealand Alps

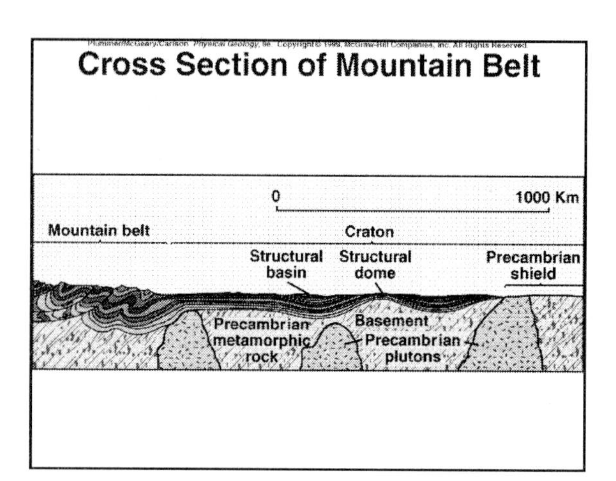
Cross Section of Mountain Belt
0
1000 Km
Mountain belt
Craton
Structural basin
Structural dome
Precambrian shield
Precambrian metamorphic rock
Basement
Precambrian plutons

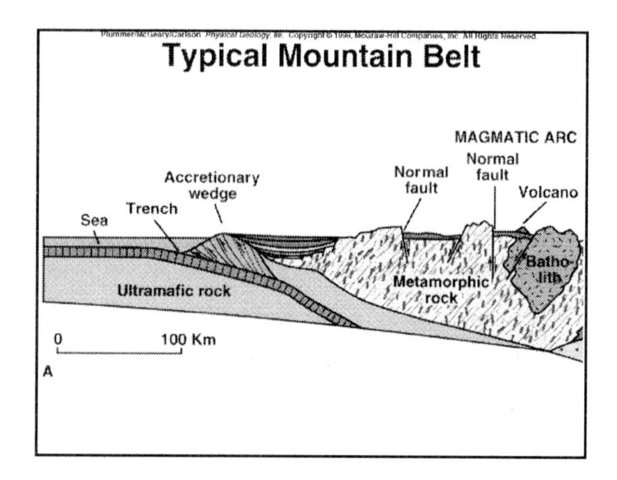
Typical Mountain Belt
MAGMATIC ARC
Normal fault
Normal fault
Volcano
Accretionary wedge
Trench
Sea
Ultramafic rock
Metamorphic rock
Batholith
0
100 Km
A

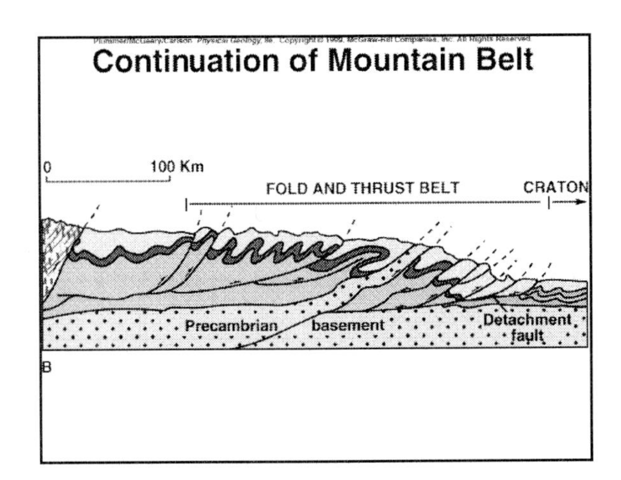
Continuation of Mountain Belt
0
100 Km
FOLD AND THRUST BELT
CRATON
Precambrian
basement
Detachment fault
B

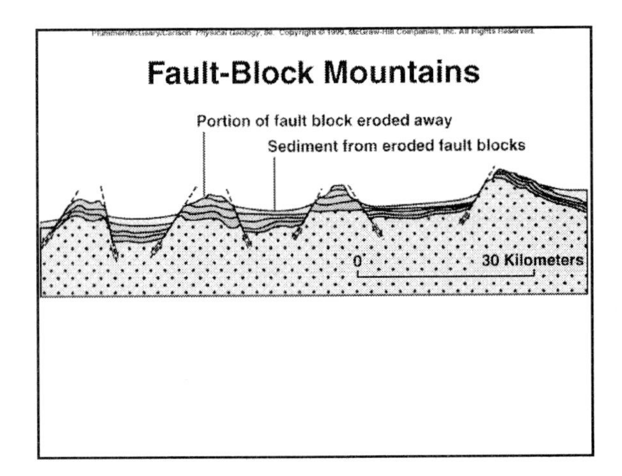
Fault-Block Mountains
Portion of fault block eroded away
Sediment from eroded fault blocks
0
30 Kilometers

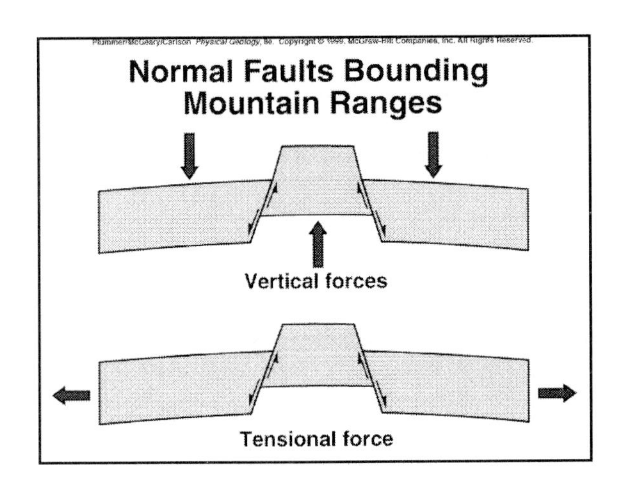
Normal Faults Bounding Mountain Ranges
Vertical forces
Tensional force

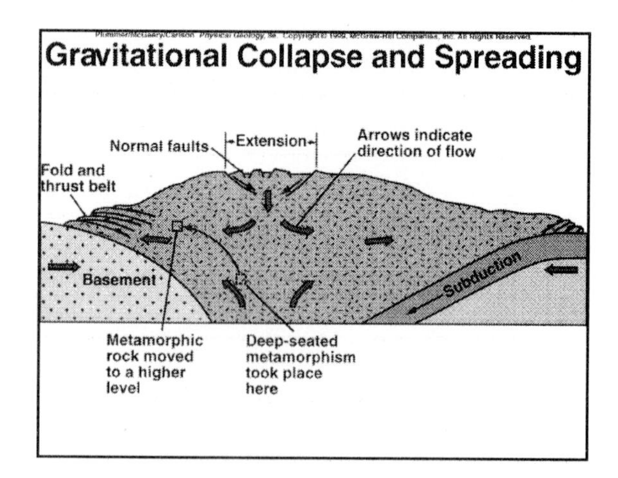
Plummer/McGeary/Carlson Physical Geology, 8e Copyright © 1999, McGraw-Hill Companies, Inc. All Rights Reserved
Gravitational Collapse and Spreading
Normal faults
Extension
Arrows indicate direction of flow
Fold and thrust belt
Basement
Subduction
Metamorphic rock moved to a higher level
Deep-seated metamorphism took place here

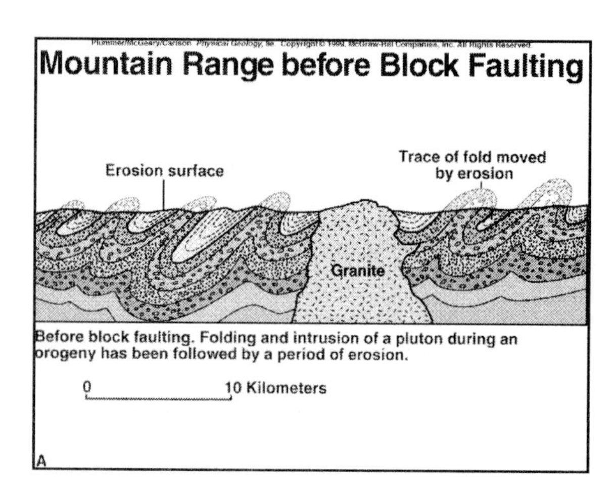
Plummer/McGeary/Carlson Physical Geology, 8e Copyright © 1999, McGraw-Hill Companies, Inc. All Rights Reserved
Mountain Range before Block Faulting
Erosion surface
Trace of fold moved by erosion
Granite
Before block faulting. Folding and intrusion of a pluton during an orogeny has been followed by a period of erosion.
0
10 Kilometers
A

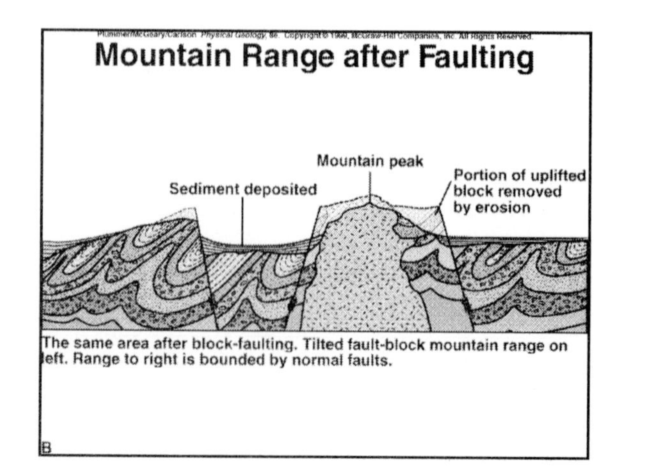
Plummer/McGeary/Carlson Physical Geology, 8e Copyright © 1999, McGraw-Hill Companies, Inc. All Rights Reserved
Mountain Range after Faulting
Mountain peak
Sediment deposited
Portion of uplifted block removed by erosion
The same area after block-faulting. Tilted fault-block mountain range on left. Range to right is bounded by normal faults.
B

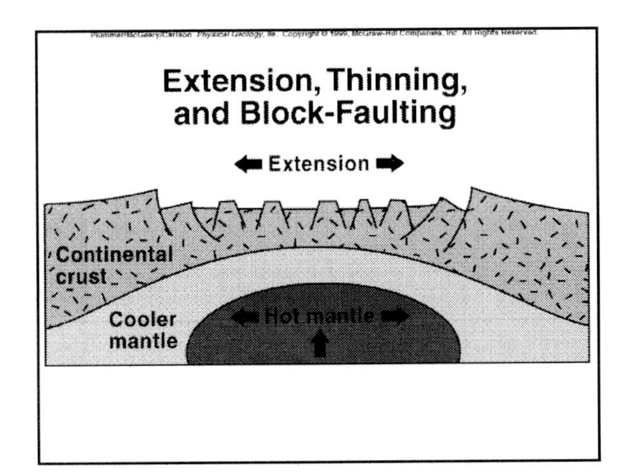
Extension, Thinning, and Block-Faulting
Extension
Continental crust
Cooler mantle
Hot mantle

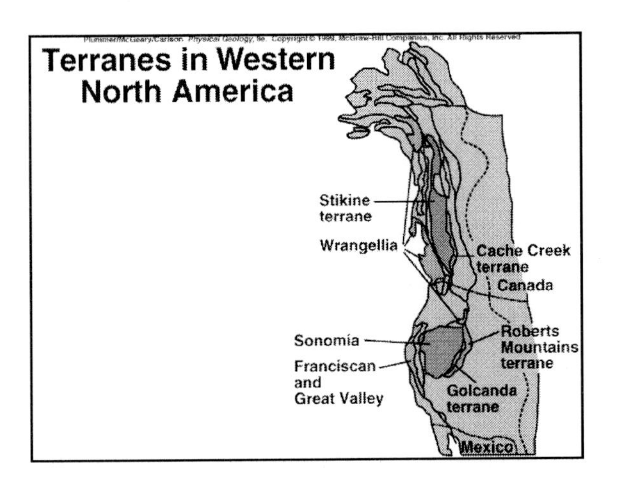
Terranes in Western North America
Stikine terrane
Wrangellia
Cache Creek terrane
Canada
Sonomia
Roberts Mountains terrane
Franciscan and Great Valley
Golcanda terrane
Mexico

Chapter 21

Physical Geology

eighth edition

Plummer/McGeary/Carlson

GEOLOGIC RESOURCES

Types of Resources

- *Geological Resources*
- Energy Resources
- Metals
- Nonmetallic Resources
- All are *nonrenewable resources*
  - Ground water an exception
- Resources vs. Reserves

Energy Use

- Oil
- Natural gas
- Coal
- Nuclear
- Hydroelectric

Oil & Natural Gas

- *Petroleum*
  - Crude oil
  - Natural gas
- Occurrence of Oil & Gas
  - Oil pool
    - Source rock
    - Reservoir rock
    - Trap- structural trap; stratigraphic trap
    - Deep enough burial
  - Oil field

Oil & Natural Gas

- Recovering oil
  - Environmental effects
- How much oil do we have left?
  - World situation
  - Outlook in United States

Heavy Crude & Oil Sands

- Heavy crude
- Oil sands (or tar sands)

Oil Shale

- Oil shale
- Problems with mining oil shale

Coal

- Varieties of coal
- Occurrence of coal
  - Strip mine
  - Appalachian fields
  - Interior fields
  - Far western fields
- Environmental effects
- Reserves & resources

Alternative Sources of Energy

- Hydroelectric power
- Geothermal power
- Solar power; wind power
- tidal power; wave power; ocean current power
- Vertical temperature differences in the sea
- Hydrogen from dissociation of water

Metals & Ores

- *Ore*

Origin of Metallic Ore Deposits

- Ores associated with igneous rocks
  - Crystal settling
  - Hydrothermal fluids
    - Contact metamorphism
    - Hydrothermal veins
    - Disseminated ore deposits
      - Porphyry copper
    - Hot springs
  - Pegmatites

Origin of Metallic Ore Deposits

- Ores formed by surface processes
  - Chemical precipitation in layers
  - Placer deposits
  - Supergene enrichment
- Metal ores and plate tectonics
  - Divergent plate boundaries
  - Convergent plate boundaries
- Mining
- Environmental effects

Origin of Metallic Ore Deposits

- Some important metals
  - Iron
  - Copper
  - Aluminum
  - Lead
  - Zinc
  - Silver
  - Gold
  - Other metals

Nonmetallic Resources

- Construction materials
  - Sand & gravel
  - Stone
  - Limestone

Nonmetallic Resources

- Fertilizers and evaporites
  - Fertilizers
    - Phosphate; nitrate; potassium compounds
  - Rock salt
  - Gypsum
  - Sulfur
- Other nonmetallics

Some Future Trends

- Ocean mining
- Metallic brines
- Improved tools & techniques
- The human perspective

A Pencil´s Mineral Resources

Zinc
Petroleum
Brass
Copper
Machinery to shape pencil
Iron
Pigment
Pigment
Clay
Graphite

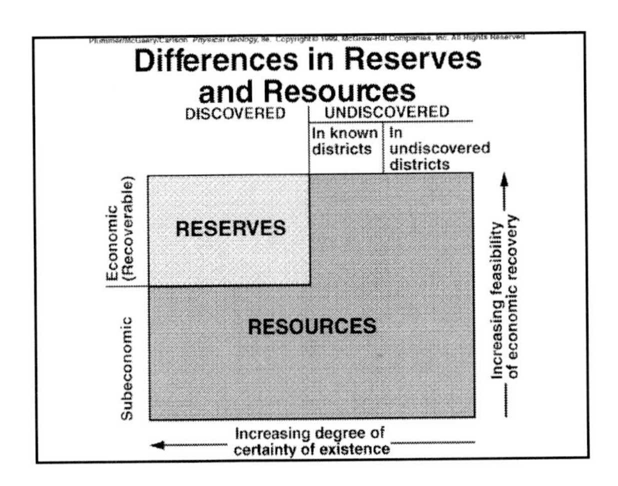
Differences in Reserves and Resources
DISCOVERED
UNDISCOVERED
In known districts
In undiscovered districts
Economic (Recoverable)
Subeconomic
RESERVES
RESOURCES
Increasing feasibility of economic recovery
Increasing degree of certainty of existence

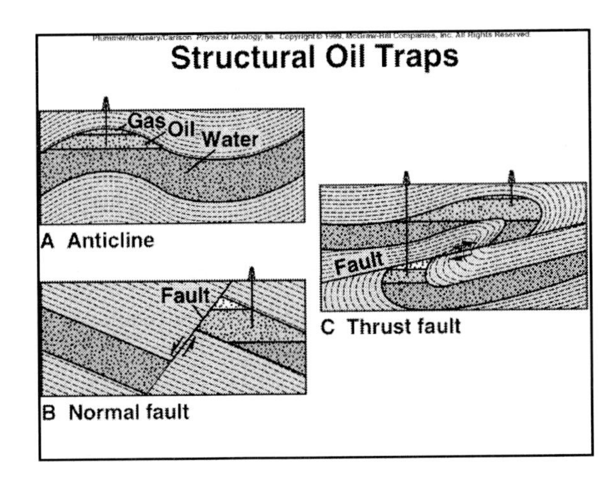
Structural Oil Traps
Gas
Oil
Water
A Anticline
Fault
B Normal fault
Fault
C Thrust fault

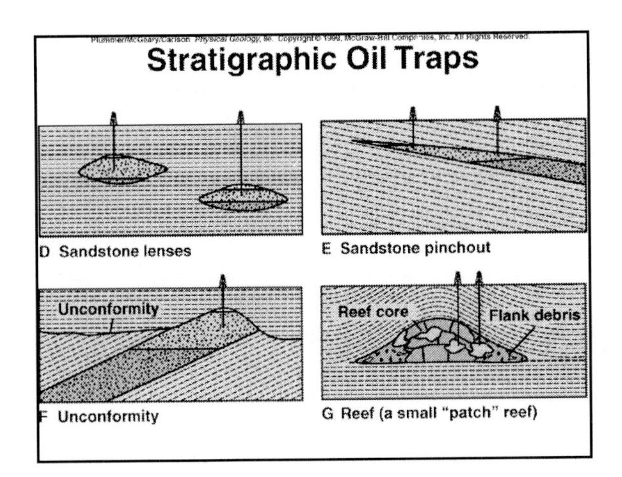
Stratigraphic Oil Traps
D Sandstone lenses
E Sandstone pinchout
Unconformity
F Unconformity
Reef core
Flank debris
G Reef (a small "patch" reef)

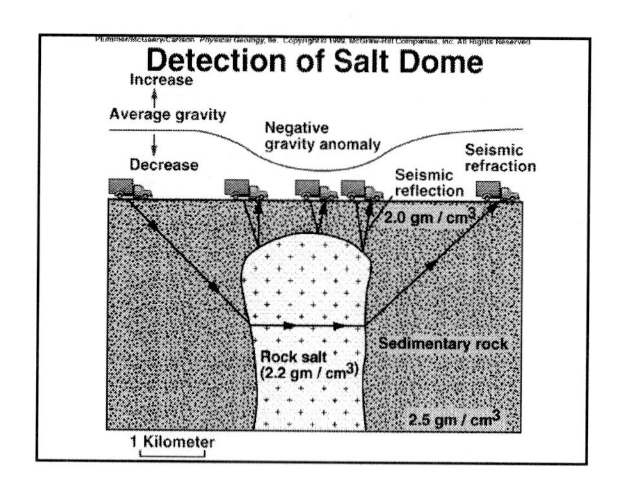
Detection of Salt Dome
Increase
Average gravity
Decrease
Negative
gravity anomaly
Seismic
refraction
Seismic
reflection
2.0 gm / cm3
Rock salt
(2.2 gm / cm3)
Sedimentary rock
2.5 gm / cm3
1 Kilometer

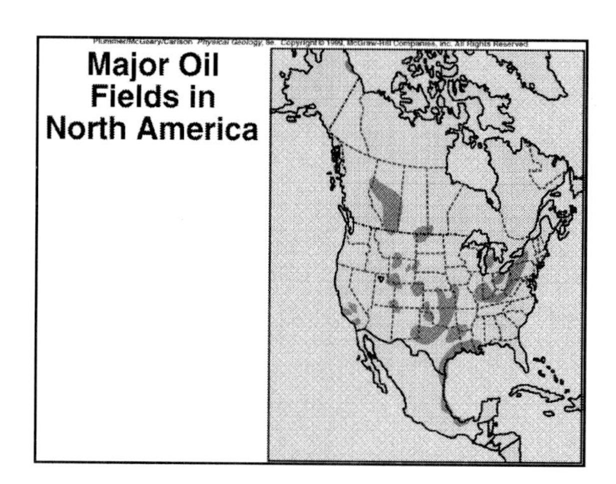
Major Oil
Fields in
North America

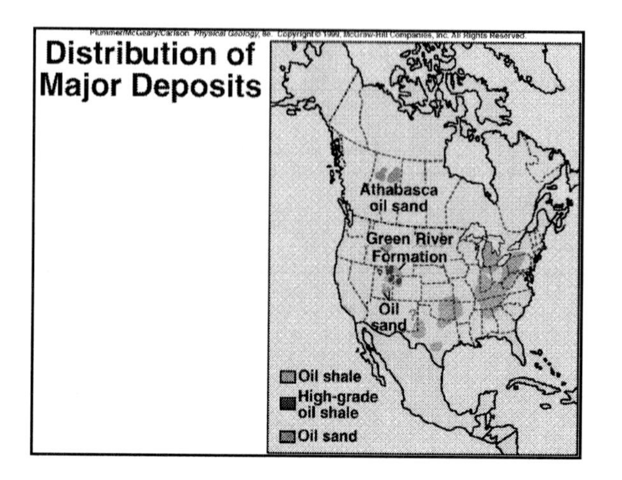
Distribution of
Major Deposits
Athabasca
oil sand
Green River
Formation
Oil
sand
Oil shale
High-grade
oil shale
Oil sand

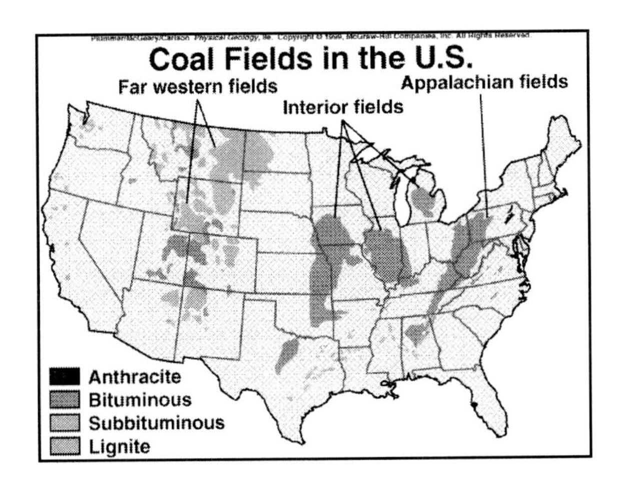
Coal Fields in the U.S.
Far western fields
Interior fields
Appalachian fields
Anthracite
Bituminous
Subbituminous
Lignite

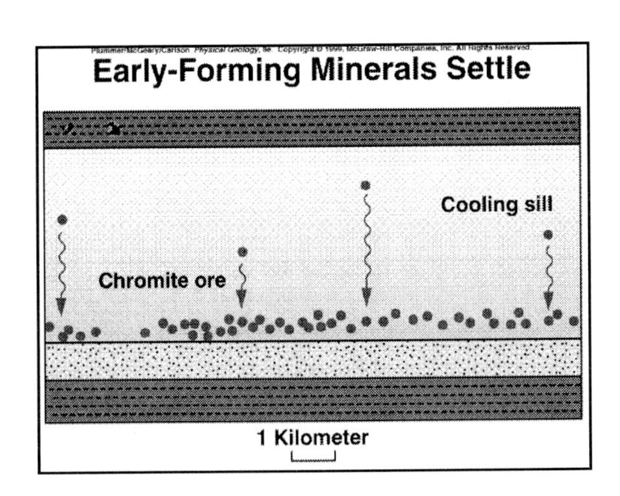
Early-Forming Minerals Settle
Cooling sill
Chromite ore
1 Kilometer

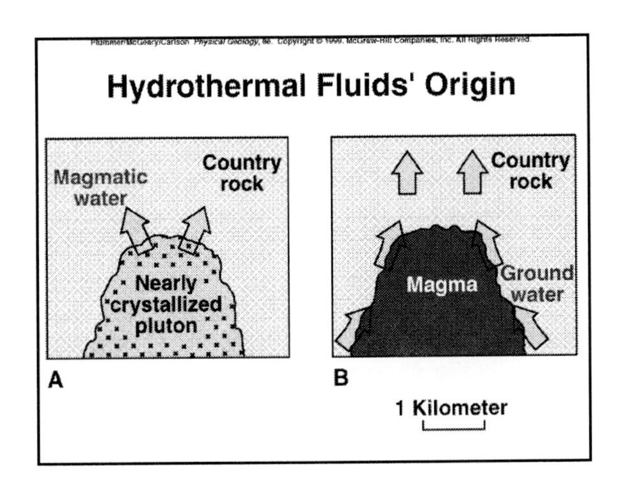
Hydrothermal Fluids' Origin
Magmatic water
Country rock
Nearly crystallized pluton
A
Country rock
Magma
Ground water
B
1 Kilometer

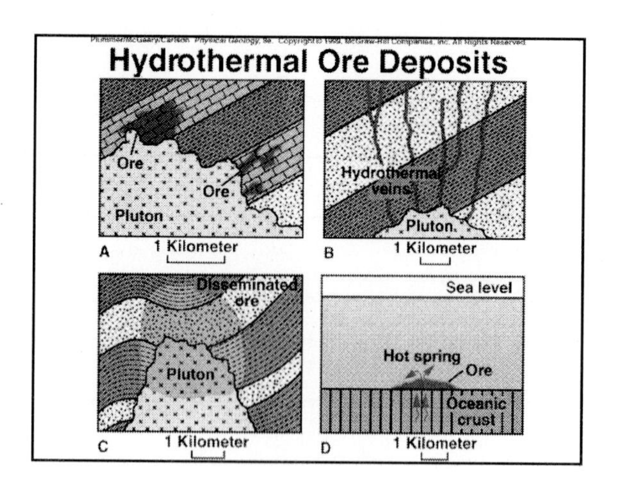
Plummer/McGeary/Carlson Physical Geology, 8e Copyright © 1999, McGraw-Hill Companies, Inc. All Rights Reserved
Hydrothermal Ore Deposits
Ore
Ore
Pluton
A
1 Kilometer
Hydrothermal veins
Pluton
B
1 Kilometer
Disseminated ore
Pluton
C
1 Kilometer
Sea level
Hot spring
Ore
Oceanic crust
D
1 Kilometer

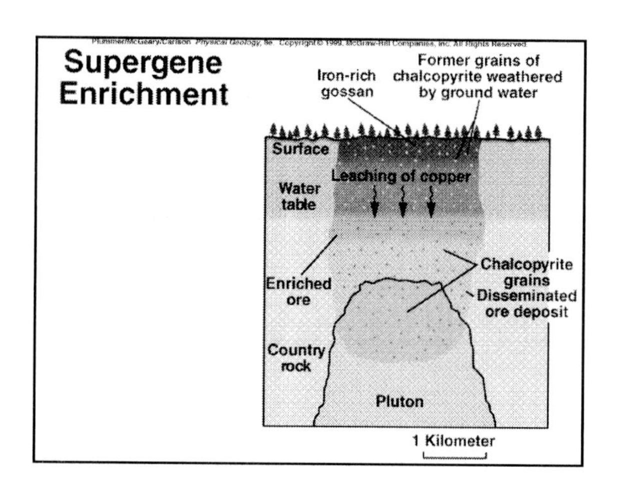
Plummer/McGeary/Carlson Physical Geology, 8e Copyright © 1999, McGraw-Hill Companies, Inc. All Rights Reserved
Supergene Enrichment
Iron-rich gossan
Former grains of chalcopyrite weathered by ground water
Surface
Leaching of copper
Water table
Enriched ore
Chalcopyrite grains
Disseminated ore deposit
Country rock
Pluton
1 Kilometer

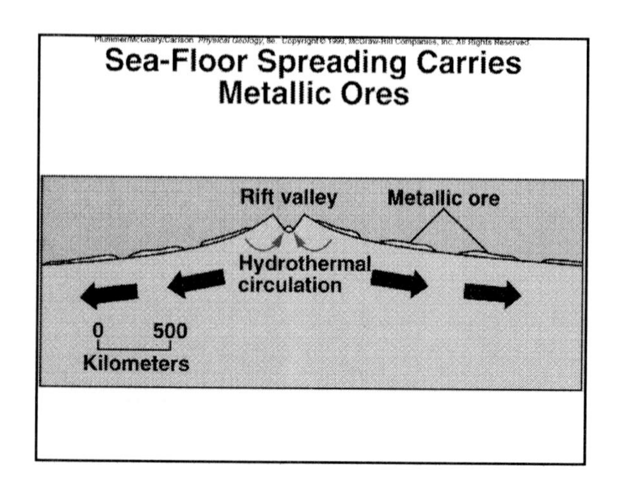
Plummer/McGeary/Carlson Physical Geology, 8e Copyright © 1999, McGraw-Hill Companies, Inc. All Rights Reserved
Sea-Floor Spreading Carries Metallic Ores
Rift valley
Metallic ore
Hydrothermal circulation
0
500
Kilometers

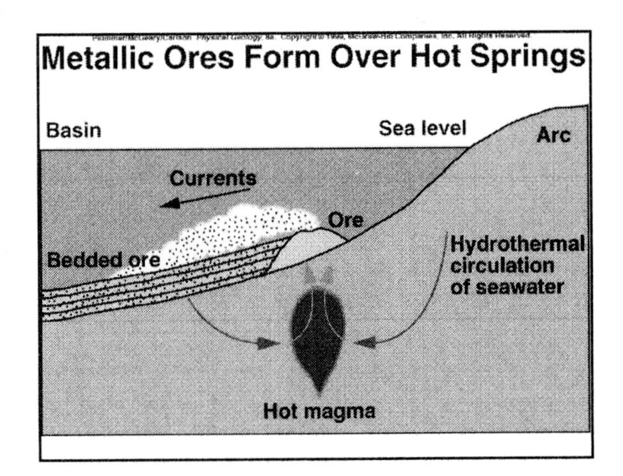
Metallic Ores Form Over Hot Springs
Basin
Sea level
Arc
Currents
Ore
Bedded ore
Hydrothermal circulation of seawater
Hot magma

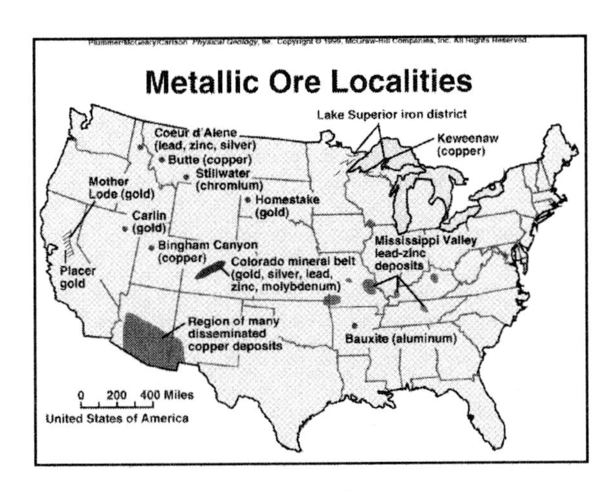
Metallic Ore Localities
Lake Superior iron district
Keweenaw (copper)
Coeur d'Alene (lead, zinc, silver)
Butte (copper)
Stillwater (chromium)
Mother Lode (gold)
Homestake (gold)
Carlin (gold)
Bingham Canyon (copper)
Colorado mineral belt (gold, silver, lead, zinc, molybdenum)
Mississippi Valley lead-zinc deposits
Placer gold
Region of many disseminated copper deposits
Bauxite (aluminum)
0 200 400 Miles
United States of America